D1174931

Oxford Chemistry Series

Oxford Chemistry Series

---

1. K. A. McLauchlan: *Magnetic resonance*
2. J. Robbins: *Ions in solution* (2): *an introduction to electrochemistry*
3. R. J. Puddephatt: *The periodic table of the elements*
4. R. A. Jackson: *Mechanism*: *an introduction to the study of organic reactions*
5. D. Whittaker: *Stereochemistry and mechanism*
6. G. Hughes: *Radiation chemistry*
7. G. Pass: *Ions in solution* (3): *inorganic properties*
8. E. B. Smith: *Basic chemical thermodynamics*
9. C. A. Coulson: *The shape and structure of molecules*
10. J. Wormald: *Diffraction methods*
11. J. Shorter: *Correlation analysis in organic chemistry*: *an introduction to linear free-energy relationships*
12. E. S. Stern (ed): *The chemist in industry* (1): *fine chemicals for polymers*
13. A. Earnshaw and T. J. Harrington: *The chemistry of the transition elements*
14. W. J. Albery: *Electrode kinetics*
16. W. S. Fyfe: *Geochemistry*
17. E. S. Stern (ed): *The chemist in industry* (2): *human health and plant protection*
18. G. C. Bond: *Heterogeneous catalysis: principles and applications*
19. R. P. H. Gasser and W. G. Richards: *Entropy and energy levels*
20. D. J. Spedding: *Air pollution*
21. P. W. Atkins: *Quanta*: *a handbook of concepts*
22. M. J. Pilling: *Reaction kinetics*
23. J. N. Bradley: *Fast reactions*

J. N. BRADLEY
University of Essex

Fast reactions

Clarendon Press · Oxford · 1975

Oxford University Press, Ely House, London W. 1

GLASGOW NEW YORK TORONTO MELBOURNE WELLINGTON
CAPE TOWN IBADAN NAIROBI DAR ES SALAAM LUSAKA ADDIS ABABA
DELHI BOMBAY CALCUTTA MADRAS KARACHI LAHORE DACCA
KUALA LUMPUR SINGAPORE HONG KONG TOKYO

ISBN 0 19 855456 7

Reproduced and printed by photolithography and bound in Great Britain at The Pitman Press, Bath

Editor's foreword

In a variety of scientific fields information of considerable significance has been discovered by extending the range of observation to high energies, small distances, large distances, and short times. This is particularly the case for the mechanism of chemical reactions, where observations which until quite recently were limited to processes occurring over a matter of seconds can now be made, in favourable cases, on a timescale of picoseconds. Even though short time-scales are important it remains necessary to have a range of techniques spanning the time-scales on which reactions advance significantly. Observation on a nanosecond time-scale, while it cannot reveal picosecond processes, might miss the slower, microsecond processes which can themselves yield important information. This book looks at the range of methods now available and under development. With an appropriate system, and an appropriate technique, we can study the most significant processes of a chemical reaction.

Several books in the series augment the material in this volume. In particular Pilling's *Reaction kinetics* (OCS 22) sets the scene for the modern study of chemical processes, and discusses some of the theories lying beneath the aims of the experiments discussed here. Albery's *Electrode kinetics* (OCS 14) looks in detail at a particular type of reaction system, but one in which fast, fundamental electron-transfer reactions are the principal interest. Fast processes can also be studied by magnetic resonance, as the present volume describes, and the background to this application will be found in McLauchlan's *Magnetic resonance* (OCS 1).

P.W.A.

Preface

OUR knowledge of molecular structure is very closely coupled to the development of theoretical understanding; however, the situation regarding dynamic processes is still primarily dictated by the experimental techniques available, though naturally theoretical studies have an ever-increasing part to play. The cause quite simply is that the relaxation of the constraint of time invariance on stationary systems makes such situations exceedingly complex. Significant advances in understanding are therefore still associated mainly with experimental innovation.

For this reason, the present volume has been sub-divided according to the type of experimental technique involved rather than in terms of classes of kinetic behaviour. As far as possible, each chapter is treated independently of the others and those teachers who wish to attach particular importance to certain specific techniques should still find this volume of value.

To compensate for the categorization by technique, illustrative examples have been selected which should provide over the complete text an overall view of recent or current interests in kinetics. In this context, the definition of kinetics has been treated quite literally and reference is made to processes such as segmental motions of polymers and vibrational energy transfer in gases.

The material covered is based on current undergraduate and postgraduate lectures in Chemistry and Chemical Physics at the University of Essex. The chapter on Shock Waves is an abbreviated version of a plenary lecture on High Temperature Chemistry presented at the Second International Symposium on Gas Kinetics held in Swansea in July 1971.

The author is greatly indebted to Penny Hardiman for preparing the typescript.

J. N. BRADLEY

Contents

1. Introduction

THE aim of this volume is to describe the methods currently available for the measurement of rates of rapid reactions and to outline the types of results which can be derived from them. It is not intended as a primer in reaction kinetics, which is dealt with quite adequately in many standard textbooks (including *Reaction Kinetics* by M. J. Pilling OCS 22). However in order to put the study of fast reactions in perspective, it is necessary first to discuss the aims of reaction kinetics.

The role of fast reactions in chemical kinetics

The subject of kinetics is concerned with an 'understanding' of the rates and mechanisms of chemical reactions: the word understanding is employed because the measurement of a rate constant and the determination of a mechanism for an isolated reaction are of only limited value. Unless one can give some answer to the question why a process displays a particular rate or reaction sequence, one would be forced to carry out long and extensive investigations of every reaction of conceivable interest. Such a prospect would soon daunt even the most ardent experimentalist.

An investigator commencing the study of a quite unknown reaction would begin by examining how the reactants are consumed and the products generated as a function of time. With care and an element of luck, he would find that the measured rates can be expressed in terms of a *rate law*, in which the differential change is described by a constant, known variously as a *rate constant*, *velocity constant*, or *velocity coefficient*, multiplied by a simple algebraic function of the concentrations of the chemical substances present. Thus, for the reaction

$$\mathrm{A + B \rightarrow C + D}$$

the rate law would take the form

$$-\frac{\mathrm{d[A]}}{\mathrm{d}t}\left\{\text{or, equally, } -\frac{\mathrm{d[B]}}{\mathrm{d}t}, \frac{\mathrm{d[C]}}{\mathrm{d}t}, \text{ or } \frac{\mathrm{d[D]}}{\mathrm{d}t}\right\} = k[\mathrm{A}]^m\,[\mathrm{B}]^n$$

The value of the constant k depends on the temperature of the system and the negative signs attached to the first two derivatives reflect the fact that [A] and [B] decrease with time.

The rate law does not normally bear any simple relation to the balanced stoichiometric equation for the overall process. If the reaction had happened to be that between hydrogen and bromine

$$H_2 + Br_2 \rightarrow 2HBr$$

the rate law would not have been the simple expression suggested by the Law of Mass Action

$$-\frac{d[H_2]}{dt} = k[H_2][Br_2]$$

but instead it would be the more involved expression

$$-\frac{d[H_2]}{dt} = \frac{k[H_2][Br_2]^{\frac{1}{2}}}{1 + k'[HBr]/[Br_2].}$$

In practice most reactions give complex rate laws valid only over limited ranges of experimental conditions.

The reason for such complex expressions is that only in the rarest circumstances do stable reactant molecules interact during a single encounter to give stable product molecules. Instead a whole sequence of steps occurs involving intermediate species of high chemical activity present in low concentration. These intermediates can be molecules that are more reactive than the original compounds, but more commonly are atoms, free radicals, ions, or electronically-excited species, with only a fleeting existence.† The complex expression above arises because the reaction mechanism comprises the following *elementary steps*:

$$\begin{aligned} Br_2 &\rightarrow 2Br \\ Br + H_2 &\rightarrow HBr + H \\ H + Br_2 &\rightarrow HBr + Br \\ H + HBr &\rightarrow H_2 + Br \\ Br + Br(+M) &\rightarrow Br_2(+M). \end{aligned}$$

An elementary step describes an actual molecular event and can be represented by a simple rate law. For example, the second of these steps is governed by the expression

$$-\frac{d[Br]}{dt} = -\frac{d[H_2]}{dt} = k[Br][H_2].$$

In order to proceed towards the understanding we are seeking, it is necessary to determine rate constants for each of the elementary steps involved in the reaction mechanism.

† See, for example, R. D. Levine and R. B. Bernstein, *Molecular reaction dynamics*, Clarendon Press, Oxford, 1974.

At this stage it is instructive to put some numbers into the expressions. At a temperature of 500 K and partial pressures of an atmosphere (100 kN m^{-2}) for each reactant, ten per cent overall conversion takes 64 seconds. Under the same experimental conditions, one-tenth conversion of the bromine atoms by the second reaction alone requires only 4.5 microseconds. This simple comparison demonstrates the importance of being able to study fast reactions.

At 'normal' concentrations, such elementary steps reach completion well within the time resolution of the human senses, which is about a second. Since reaction times of second-order processes depend on concentration, it may be possible in principle to make the measurements at extremely low concentration and then scale them up by means of the appropriate rate law. In practice, this procedure rarely affords a solution since one merely succeeds in replacing a time-response problem by a sensitivity problem. In the example above the second reaction would be measurable on a human timescale if the hydrogen pressure were about 10^{-3} Torr (130 N m^{-2}) but under such conditions the mechanism would be totally different, the bromine atoms simply disappearing at the vessel walls.

The reaction quoted is not particularly fast for an elementary step. The neutralization reaction

$$H^+ + OH^- \rightarrow H_2O$$

at ionic concentrations equal to those of the reactant gases above would take 3×10^{-11}s for one-tenth reaction and even at a more typical pH of 7 the corresponding reaction time is 7 μs.

The study of fast reactions embraces most obvious elementary steps, even though these steps commonly combine together to give a slow overall process because the reactive species are present in such small quantities. Under certain conditions, such as occur in combustion and explosion, the overall process may itself be rapid and demand the use of sophisticated techniques. However, sufficient rationale is provided for the study of fast reactions by equating them with elementary steps.

The rate constant k is commonly expressed in the Arrhenius form, $k = A\exp(-E/RT)$, where the experimental activation energy E describes the temperature dependence of the rate constant. This relation is strictly empirical; thecretical treatments of reaction rates predict rate laws of the form $A' \, T^n \exp(-E'/RT)$ and it is only because the temperature dependence due to the exponential term normally swamps any variation with temperature of the pre-exponential term that, for practical purposes, the latter is incorporated in the activation energy.

The existence of the strong temperature dependence means that reactions become very much slower as the temperature is reduced; a reaction with an activation energy of 120 kJ mol^{-1} has a rate which is ten orders of magnitude slower at room temperature than at 300°C. This suggests that one practical way of bringing fast reactions into a range where they can be measured by conventional techniques is to carry out the investigation at very low temperatures. The principal shortcoming of this approach lies in ensuring that the mechanism of reaction remains unaffected as the temperature is reduced; in those cases where such a guarantee can be provided the technique is entirely satisfactory.

The exponential temperature dependence of the rate constant illustrates another important feature of chemical reactions. This arises from the Boltzmann principle that the distribution of molecules among the available energy levels is governed by the expression $n_j \propto n_o \exp(-\epsilon_j/kT)$. Even in an elementary reaction step the molecules are distributed over an enormous number of levels and only a small fraction, lying in a fairly specific range of energies, actually undergo reaction. This causes two problems: first, the equilibrium distribution may be distorted by the occurrence of reaction and, secondly, the rate constant is very much a bulk property of the system averaged over all species present and is only indirectly related to those molecules actually undergoing reaction. The former does not necessarily invalidate a rate measurement although it may limit its range of applicability.

The ideal kinetic experiment to circumvent the latter would consist of 'firing' an isolated reactant molecule at an isolated target molecule under precisely-defined conditions and following the outcome. Ideal as this approach may seem, it is closely approximated by the crossed molecular beam approach (Chapter 2) in which two streams of molecules are allowed to react very much in this way.

Reaction cross-sections

A molecule A which enters a cell of unit dimensions containing n_B molecules of type B may either pass through unaffected or strike one of the B molecules and be scattered out of the cell in a different direction. The probability of the latter occurring depends on the ratio between the total cross-sectional area presented by the B molecules and the area of the cell-face (n_B is assumed to be sufficiently small that both 'shadowing' of one molecule by another and multiple collisions can be neglected). A collision of hard-sphere molecules occurs when the centres of the two molecules reach a separation equal to the sum of the two radii $r_A + r_B$. The same result is achieved if molecule A is treated as a point and molecule B assumes a radius equal to $(r_A + r_B)$. In the case of the unit cell the probability of collision is simply $n_B.\pi(r_A + r_B)^2$. If instead of a single molecule A, a stream of molecules of density n_A enters at a velocity v, the number of molecules scattered in unit time is equal to the product of the total flux $n_A \cdot v$ and the probability

$$I = n_A n_B v \pi(r_A + r_B)^2.$$

The intensity is thus determined by two concentrations, a relative velocity, and an area. Since molecules are rarely spherical it is preferable to replace the term $\pi(r_A + r_B)^2$ by a *cross-section* σ. As the only process so far considered has been elastic scattering, σ corresponds to the *total collision cross-section*; however, this concept can easily be extended to a wider range of processes. Firstly not all molecules which undergo collision will be scattered through the same angle and the intensity of molecules scattered through a particular angle θ proves to be of interest. The intensity, $I(\theta)\,\mathrm{d}\theta$, of particules scattered into a small range of angles $\mathrm{d}\theta$ surrounding the angle θ (see Fig. 1.1) is given by

$$I(\theta)\,\mathrm{d}\theta = n_A n_B v \sigma_d(\theta, v)\,\mathrm{d}\theta$$

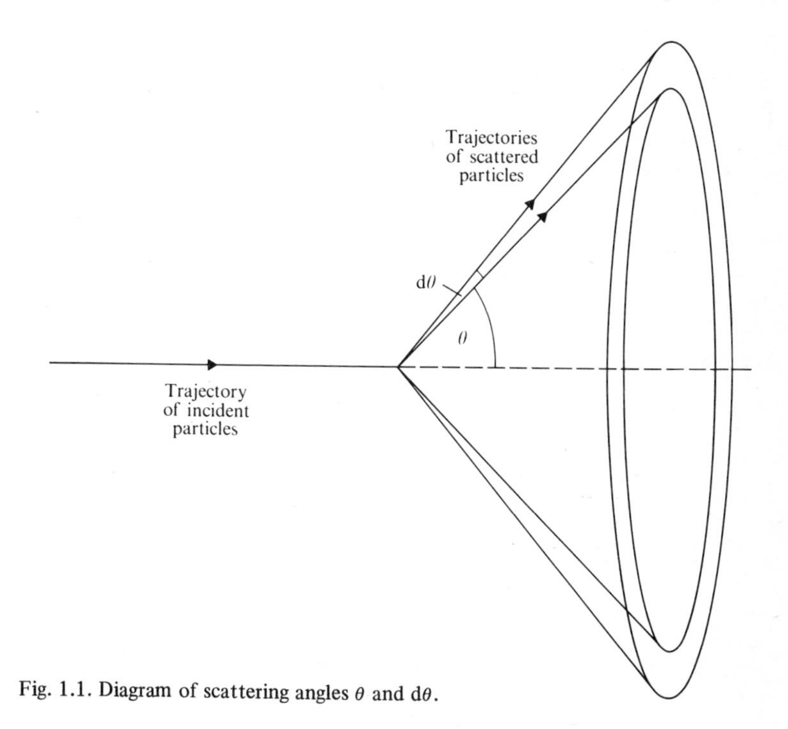

Fig. 1.1. Diagram of scattering angles θ and $\mathrm{d}\theta$.

σ_d now becomes a function of v as well as of θ. The total probability is the sum of these probabilities σ_d over all possible angles: for this reason $\sigma_d(\theta, v)$ is termed the *differential scattering cross-section.* The cross-section refers to the intensity of particles scattered in a given direction although it could just as easily have been restricted to particles which simultaneously underwent a chemical change. σ_d would then be replaced by σ_{dr}, the *differential reaction cross-section.* Since reaction is unlikely to occur on every collision, even for high values of v, σ_{dr} is expected to be considerably less than σ_d. However, molecules are not solid bodies like billiard balls and, in certain instances, the cross section for reaction may be higher than for elastic scattering.

The concept of a cross-section proves very useful in kinetics but provides a continuing source of confusion to the beginner. One way of

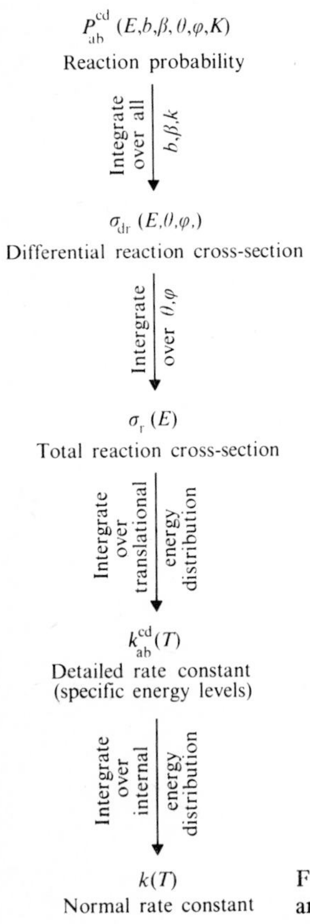

$P_{ab}^{cd}\ (E,b,\beta,\theta,\varphi,K)$

Reaction probability

↓ Integrate over all b,β,k

$\sigma_{dr}\ (E,\theta,\varphi,)$

Differential reaction cross-section

↓ Intergrate over θ,φ

$\sigma_r\ (E)$

Total reaction cross-section

↓ Intergrate over translational energy distribution

$k_{ab}^{cd}(T)$

Detailed rate constant
(specific energy levels)

↓ Intergrate over internal energy distribution

$k(T)$

Normal rate constant

Fig. 1.2. Relation between reaction probability and conventional rate constant.

overcoming the problem is to regard the cross-section as indicating the size of the equivalent hard-spheres which would lead to the same rate as the real molecules, if they underwent the process under consideration on every collision. The cross-section is therefore equivalent to a rate constant or a reaction probability; if the reaction cross-section has a numerical value equal to $\frac{1}{100}$th of the collision cross-section, normally obtained from viscosity measurements, then it signifies that one in every hundred collisions leads to reaction.

At this stage it is profitable to state the most fundamental type of information one can define for a particular chemical reaction. In a reaction between A and B to give C and D, where the species possess internal energy states (rotational, vibrational, and electronic) a, b, c, and d, it is possible to define a reaction probability $P_{\mathrm{ab}}^{\mathrm{cd}}(E, b, \beta, \theta, \phi, \kappa)$ which is a function of E, the relative kinetic energy, b the *impact parameter*, which measures the 'closeness' of the collision (see Fig. 1.3), β the orientation of the plane in which it occurs, θ, ϕ the scattering angles, and κ the collision complex involved. By successive integration (see Fig. 1.2) this quantity can be related to the cross-section defined above and to the conventional rate constant.

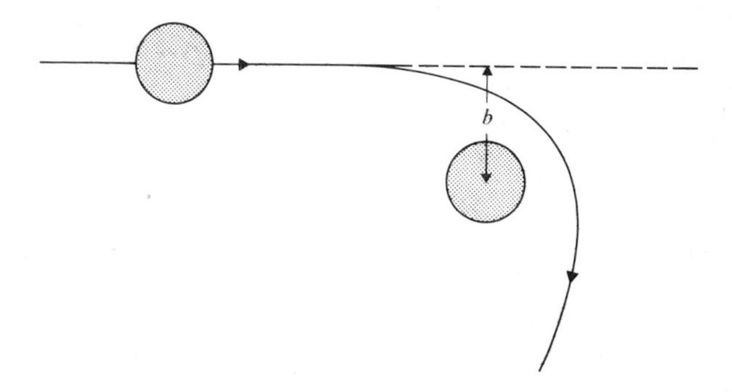

Fig. 1.3. Significance of impact parameter b.

Although the quantity $P_{\mathrm{ab}}^{\mathrm{cd}}$ cannot be determined experimentally, the differential cross section for molecules in specified energy states can be measured for a number of reactions by crossed molecular beam techniques, as we shall see. Kinetic measurements can thus be conducted at two quite different levels of sophistication, one of which provides a conventional rate constant and the other which gives the reaction probability for more closely specified conditions.

Classification of experimental techniques

Leaving aside the crossed molecular beam technique it is possible to enumerate certain principles involved in the measurement of rapid reaction rates. The problem is not simply one of timing; the species of particular interest are usually there in low amounts in the presence of other substances at higher concentrations. Additional requirements of sensitivity and selectivity are therefore imposed on the experimental techniques.

The first and perhaps clearest distinction one can make is between those techniques which operate in 'real-time', which necessarily means below a second, and those which circumvent the timing problem by replacing it with an alternative measurement. The relation between the various approaches is depicted schematically in Fig. 1.4.

With real-time systems the first requirement is a detection technique which has an adequate time response. Photoelectric devices are particularly suitable and the added advantage of high selectivity associated with optical spectroscopy means that the majority of real-time systems employ spectroscopic detection. The best combination of sensitivity, selectivity, and time response is usually obtained in the ultraviolet and visible regions of the spectrum although infrared and X-ray systems have found restricted use. A further advantage of optical spectroscopy is that interference with the reaction system is minimal.

The next most valuable analytical tool is the mass spectrometer. This device, used in conjunction with an electron multiplier-detector, has the same time response as an optical spectrometer since the detector operates on identical principles. The selectivity of a mass spectrometer is somewhat better in that it responds to all species present and not just to those with particular absorption characteristics. This advantage brings with it two attendant drawbacks, namely that the species under investigation must not fall below a certain fraction of the total and the high electron energies employed to achieve adequate sensitivity produce overlapping cracking patterns. The mass spectrometer also demands the physical introduction of the sample; this leads simultaneously to interference with the system and to the problem of extracting a truly representative sample from the reactor.

Beyond these techniques one has to forego the advantage of selectivity and accept methods which monitor a property of the system only indirectly associated with the concentration of a specific component, such as pressure, density, or electrical conductivity.

An alternative approach sometimes available is to quench the reaction very rapidly and revert to more conventional methods of analysis. This 'single-pulse' technique remains a real-time method and to some extent

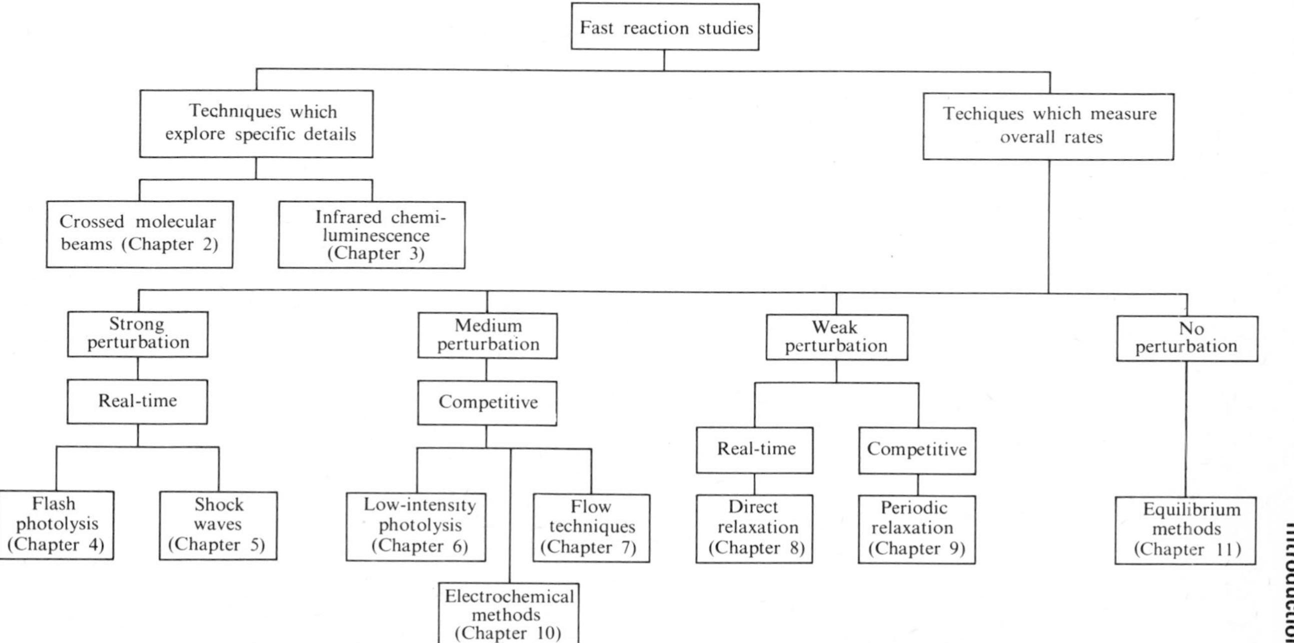

Fig. 1.4. Classification of experimental techniques.

combines the best of both worlds but, unfortunately, rapid quenching can only rarely be accomplished.

If one is operating in real-time, initiation of reaction becomes critical since it must take place within a time interval substantially shorter than the measurement time. The 'classical' way of initiating reaction by mixing substances which react spontaneously cannot be completed in less than a millisecond, too long a period for many fast reactions. The reagents must therefore be mixed beforehand and the reaction initiated by a rapid alteration of the physical conditions.

A further sub-division can be introduced depending on the magnitude of the perturbation employed. If the perturbation is small, a system, originally at equilibrium, will be displaced from it and will then return or *relax*, the effect of the disturbance (for example an increase in energy) being dissipated to the surroundings. With medium and large perturbations the displacement is considerable and relaxation is towards a different equilibrium position. Large perturbations may be employed not merely to initiate reaction but also to transfer the system to a combination of physical conditions not readily accessible otherwise. For example, the shock-tube technique, described in Chapter 5, raises the gas to a temperature of several thousand degrees in a fraction of a microsecond without any energy transfer from the container walls.

The perturbation normally increases the total energy of the system although which parameter is initially affected has a controlling effect on the subsequent relaxation. The shock-tube method already referred to first increases the translational energy of the gas and this is then redistributed among the other energy modes. Flash photolysis, on the other hand, initially populates the electronic levels and the energy is subsequently degraded, with translation often the last mode to relax.

In the liquid phase, the disturbance may be introduced as pressure, temperature, or even electrical energy.

The difficulty of working in real-time can be avoided by using a continuous or repetitive form of initiation and allowing the chemical event to compete against a suitable physical process so that a steady-state is established. Provided the rate of the physical process is known the chemical rate is obtained as a comparative measurement. The principle may be appreciated by considering the simple flow system depicted in Fig. 1.5. Two species which react spontaneously are introduced at high velocity through inlets A and B. Apart from a short period of turbulent mixing, a smooth flow occurs along the tube and the concentrations decrease as a function of distance. Provided the reservoirs of reactants are of adequate size a steady state is established and rate determinations may be accomplished at leisure by monitoring

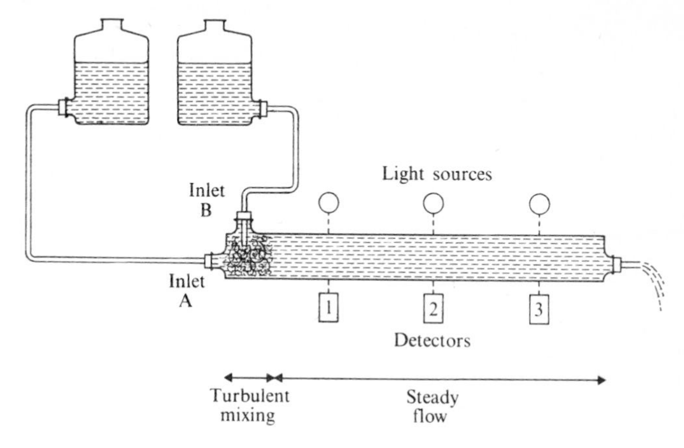

Fig. 1.5. Simple flow system.

the reactant concentrations as a function of position along the tube. The time dependence is thus replaced by a distance dependence, the two being related by the flow velocity.

The same principle is employed when repetitive initiation is used although the manner in which the data are obtained is somewhat different. A system which is subjected to a slowly-changing perturbation will adjust to it and follow it closely. At the other extreme, with a rapidly-oscillating perturbation, the system will be unable to keep up with the change and will be largely unaffected by it. Between the two extremes, where the rate of relaxation is comparable to the rate of change of the perturbation, the system absorbs energy and the frequency of the perturbation provides a measure of the relaxation rate.

Finally certain properties exist which are characteristic of the rate processes involved in a dynamic equilibrium, notably the line-broadening of spectral transitions associated with the mean lifetimes of individual states. The system does not suffer an external perturbation and an approach based upon these properties may be classified as an equilibrium method.

The characteristic reaction times to which the various techniques apply are listed in Table 1.1; the corresponding rate constants depend on the reaction order and, for bimolecular reactions, on the concentrations employed.

General features of fast reactions

It is not possible to regard fast reactions simply as 'normal' reactions which happen to occur more quickly. Bimolecular gas phase reactions

TABLE 1.1
Reaction times observable by various techniques

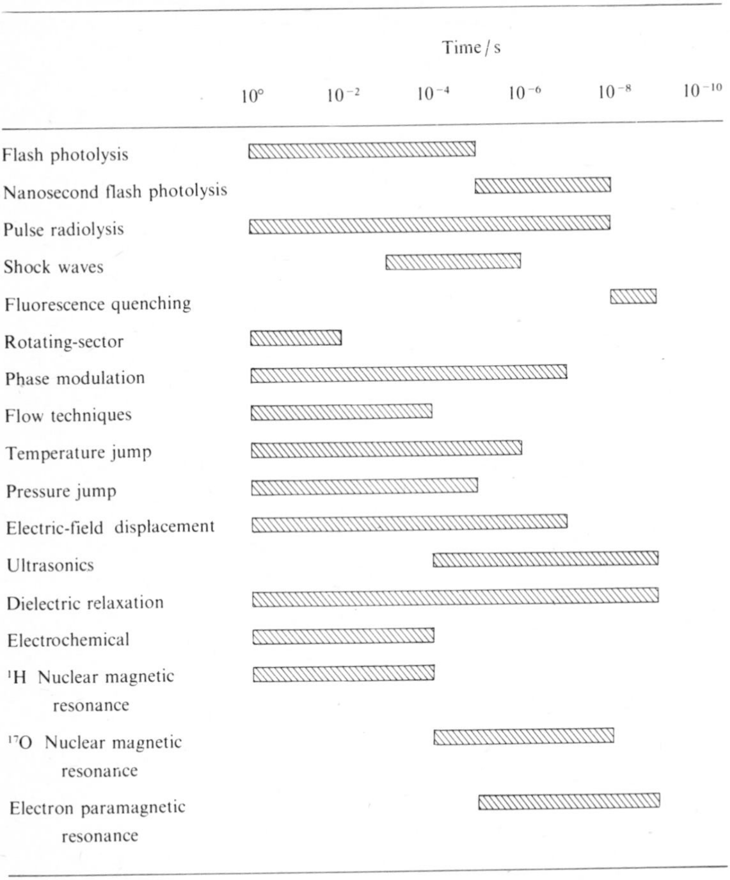

Technique	Time / s (10^{0}, 10^{-2}, 10^{-4}, 10^{-6}, 10^{-8}, 10^{-10})
Flash photolysis	10^{0} – 10^{-5}
Nanosecond flash photolysis	10^{-5} – 10^{-8}
Pulse radiolysis	10^{0} – 10^{-8}
Shock waves	10^{-3} – 10^{-6}
Fluorescence quenching	10^{-8} – 10^{-9}
Rotating-sector	10^{0} – 10^{-2}
Phase modulation	10^{0} – 10^{-7}
Flow techniques	10^{0} – 10^{-4}
Temperature jump	10^{0} – 10^{-6}
Pressure jump	10^{0} – 10^{-5}
Electric-field displacement	10^{0} – 10^{-7}
Ultrasonics	10^{-4} – 10^{-9}
Dielectric relaxation	10^{0} – 10^{-9}
Electrochemical	10^{0} – 10^{-4}
^{1}H Nuclear magnetic resonance	10^{0} – 10^{-4}
^{17}O Nuclear magnetic resonance	10^{-4} – 10^{-8}
Electron paramagnetic resonance	10^{-5} – 10^{-9}

† Because the measurable extent of reaction may depend on temperature and concentration, a narrow range of times does not of itself imply a lower versatility.

are rapid when the energy requirement is low and the activation energy is small. Such behaviour is characteristic of highly reactive intermediates and the kinetics of such species are related to their overall chemistry. Although some information on these reactions can often be gleaned by relative rate studies, direct measurements are needed to complete the picture. It turns out that traditional ideas on rate behaviour have to be revised for these highly reactive entities. For example, the addition of oxygen and sulphur atoms to olefins has a low activation energy which, with some molecules, actually becomes negative. Reactions of hydroxyl radicals also display low or zero activation energies and it appears that an Arrhenius relation may not represent such data satisfactorily.

A further consequence is that the speed of such reactions eventually becomes comparable to the rate of energy transfer. It has been appreciated for some time that the overall rate of a unimolecular reaction at high temperature or low pressure is governed by the rate of energy transfer rather than by the chemical rearrangement. It now becomes possible to investigate how reaction rates are affected by different forms of energy in the reactants and to discover in which modes the energy released will appear. A recent investigation shows that the rate of the reaction $H_2 + CN$ is enhanced if the CN radical is vibrationally excited; in contrast, the corresponding $O_2 + CN$ reaction is retarded. Fast reaction techniques are revealing more and more examples in which the reaction behaviour is affected by the actual energy distribution, and such information eventually will lead to a more detailed understanding of reaction dynamics at a molecular level.

When the reaction takes place in solution, the dominant question concerns the role which diffusion plays in determining the kinetics. In a bimolecular reaction, the highest possible rate is presumably the frequency with which the reactant molecules are able to diffuse together through the solvent. The flux F is governed by Fick's law, which makes the flux directly proportional to the concentration gradient:

$$\Gamma = -D\frac{\mathrm{d}n}{\mathrm{d}x}.$$

At the maximum rate a concentration gradient will be established in which the concentration of a reactant B varies from its bulk value n_B, at infinite separation, down to zero, at the surface of a sphere of radius $r_A + r_B$ around the other reactant A for, on this spherical surface, the reaction removes A and B from the solution. Integrating over all internuclear separations gives the limiting diffusion rate as $4\pi(D_A + D_B)(r_A + r_B)n_A n_B$. The diffusion coefficients D_A and D_B are conveniently replaced by viscosities using the Stokes-Einstein formula $D = kT/6\pi r\eta$ to give the following expression for a diffusion controlled rate constant

$$k_D = \frac{2RT}{3\eta} \cdot \frac{(r_A + r_B)^2}{r_A r_B}.$$

This elementary treatment is adequate to demonstrate that k_D assumes a value of about 10^{10} l mol^{-1} s^{-1} for aqueous solutions (much lower for viscous solvents) compared with the corresponding collision rate which is a factor of ten greater at 10^{11} l mol^{-1} s^{-1}.

If the chemical rearrangement is sufficiently rapid, the observed rate is independent of the chemistry and is related entirely to the properties of the solvent. Any temperature dependence will refer to the temperature-coefficient of the viscosity rather than to the barrier opposing reaction. As a general rule diffusion control becomes important if the energy barrier for the chemical process falls below 20 kJ mol^{-1}.

The study of fast reactions in solids has been excluded from the text deliberately. There are two reasons for this; first, comparable kinetic techniques are not available and, in consequence, considerably less is known about such reactions. Secondly, and perhaps more important, rapid processes in solids are determined more by physical factors, such as diffusion, heat evolution, and change of phase, than by chemical reactions involving rearrangements within molecules. Such interests fall more naturally into the realm of solid-state physics.

2. Crossed molecular beams

IN the previous chapter, a method of describing rate processes in terms of fundamental parameters was discussed and a notional experiment invoked to explain the significance of the term 'reaction cross-section'. In this ideal kinetic experiment a molecule is generated in a prescribed internal energy state and allowed to interact in a specified manner with a second molecule also in a pre-determined energy state. The nature, energy, and orientation of the resulting species, reacted or unreacted, is then measured. By repeating the experiment many times under closely similar conditions a probability or cross-section would be obtained for each possible outcome. Although such an idealized experiment cannot be undertaken, the crossed molecular beam technique comes very close.

A *molecular beam* is a stream of molecules travelling essentially in one direction, whose density and spread of velocities is such that collisions between the beam molecules are extremely rare. Provided the background pressure in the containing vessel is maintained at a low level, the velocities and internal energy states of the beam molecules will not change. The principle of the experiment is to allow two such beams to cross and to monitor the result by means of an appropriate detector, as shown in Fig. 2.1. The capabilities of the technique depend mainly on

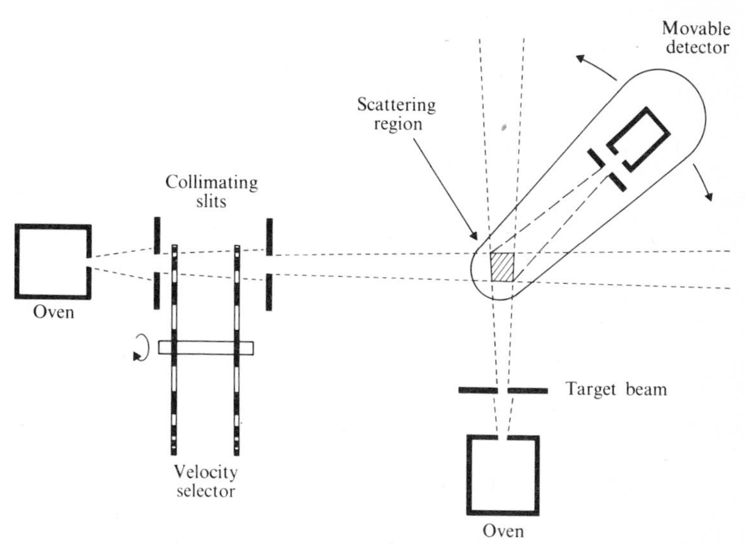

Fig. 2.1. Crossed molecular beam apparatus.

the extent to which the various properties of the beam can be specified and on the sensitivity of the detector to the separate measurements required.

Molecular interactions can lead to three types of scattering process:

(1) *Elastic scattering*, in which translational energy and linear momentum are conserved throughout.

(2) *Inelastic scattering*, in which internal energy is exchanged with translational energy.

(3) *Reactive scattering*, in which chemical reaction occurs and atoms are transferred between collision partners.

The crossed-beam technique permits the study of all three, although we shall be concerned here only with reactive scattering.

Formation of molecular beams

Effusive sources

The commonest way of generating a molecular beam is to retain the reactant in a small container or oven in one side of which is situated a small orifice. Provided the diameter is less than the mean free path of the vapour in the oven, the molecules pass independently of each other, or *effuse*, through the orifice. The intensity of the beam decreases with the cosine of the angle from the axis and it is necessary to use a series of collimating apertures to select only molecules moving within a limited solid angle.

The velocity spread matches the usual Maxwellian distribution for the oven temperature and a *velocity selector* is employed to isolate the molecules whose translational energies fall within the desired range. The usual device employed for velocity selection, very similar to that used by Fizeau to measure the velocity of light, consists simply of a pair of toothed discs rotating on a common axis. Only those molecules with the appropriate velocity will pass through a slot on the first disc and encounter a corresponding slot on the second, the remainder being blocked by the solid portions of the discs. The same technique can be used to analyze the velocity (kinetic energy) distribution of the products.

A typical beam intensity, before velocity selection, corresponds to a density of 10^{-6} mol m^{-3} or a pressure of 13 μN m^{-2} (10^{-7} Torr). Although such effusive beams have been used most frequently, the subsequent collimation and velocity selection mean that the beam intensity is further reduced and hence severe demands are imposed on the sensitivity of the detector.

Supersonic sources

An alternative source employs an orifice or nozzle whose effective dimensions are greater than the mean free path. With such a system, the molecules are no longer collision-free but behave instead as a continuous fluid expanding adiabatically into a vacuum. In this type of flow the fluid initially converges until at the 'throat' it becomes sonic i.e. its velocity equals the local sound velocity. At this stage the flow starts to diverge and becomes supersonic (Fig. 2.2). During the

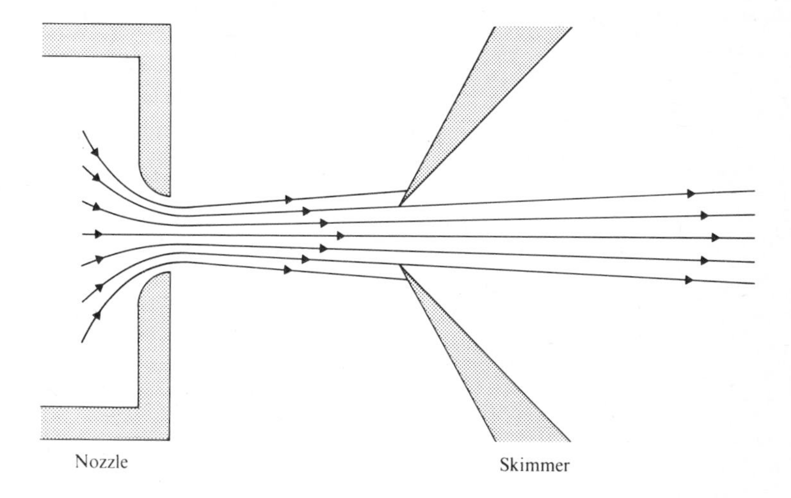

Fig. 2.2. Flow through supersonic nozzle.

expansion, the particle density falls and, as long as the gas behaves as a continuous fluid, the temperature also falls adiabatically, the random motions of the molecules which are responsible for the temperature energy being converted into ordered flow motion of the gas. Eventually the density falls to the point where the collison rate is inadequate for this exchange process to occur. At this stage, the motions of the particles no longer change so that the random or transverse motions, and hence the temperature, become 'frozen'. Until this 'freezing' occurs random translational motions within the source are converted into ordered motions along the axis, a typical 'transverse' temperature being $\frac{1}{20}$th of the original source temperature. The combination of higher source pressures and directed motion, plus the considerably reduced velocity spread, enable intensities greater by two or three orders of magnitude to be obtained. The reduction in velocity spread may even allow velocity selection to be dispensed with altogether. The only serious penalty of the supersonic source is the extremely high pumping speeds necessary to preserve the integrity of the jet and dispose of the much greater quantity of material passing through the orifice. The aerodynamics of supersonic nozzle flows has only recently been understood, and many of the earlier attempts failed to utilize the full potential of the device. It seems reasonable to predict that, in the future, supersonic sources will largely replace effusive sources.

An additional limitation of the effusive source is that the average energy, which is determined by the oven temperature, is relatively low and in order to obtain molecules of sufficient energy to undergo reaction, it is necessary to select from the low intensities in the Maxwell-Boltzmann 'tail'. Although the supersonic nozzle does not overcome this problem directly, higher energies can be achieved by 'seeding' a beam of light atoms with a heavier species. Because the expansion is a continuum process, the trace quantity of the heavier species is accelerated to the same velocity as the lighter gas and hence reaches kinetic energies much larger than those in the oven. At the same time, because of the properties of aerodynamic flow, the heavy species tend to concentrate along the centre-line. (These properties have been used to separate the carrier gas from the outflow of a gas chromatograph prior to entering a mass spectrometer detector.)

Charge-exchange sources

Although the seeded supersonic beam has afforded a way of producing energies as high as 900 kJ mol^{-1} (215 kcal mol^{-1}), there is a considerable demand for even greater values. The most common method is to form a beam of ions using an electron impact source, accelerate them to the required energy in an electric field, and then to neutralize their charge, but not their velocity, by permitting electron-exchange with a crossed beam of neutral molecules. This produces species in the 10 - 1000 eV (1 - 100 MJ mol^{-1}) range. Unfortunately electrostatic repulsion makes it difficult to obtain satisfactory beams at lower energies and a rather awkward gap exists between the various techniques.

Detection of molecular beams

The greatest limitation on the use of the crossed beam technique is the lack of suitable detectors which can provide both the required sensitivity and, in fact, only two types of detector have been of much value.

Surface ionization detector and selectivity

When an alkali metal atom falls on a heated wire of appropriate *work function*, the atom is ionized and may be collected on a suitably-situated electrode. A similar process occurs when a compound of the metal falls on the wire, although the relative efficiency of the two processes depends critically on the nature of the surface. Thus by comparing the signals obtained from two different wires it is possible to measure both the intensity of a scattered beam of atoms and the corresponding yield of product molecules. The first detector of this type employed one tungsten and one platinum wire though it was subsequently discovered that similar results can be achieved with two wires of the same material which had received different pre-treatment.

This detector has been of enormous value in studying the reactions of potassium, rubidium, and caesium and can also be employed with beams of lithium and sodium. However, since the sensitivity of the device depends on the material under investigation having a low ionization potential, it is unlikely that its use can be extended beyond the alkali metals.

Universal, or mass spectrometer, detector

A detector consisting of an electron-bombardment ion source combined with a quadrupole mass filter is certainly the most useful device so far developed. A quadrupole mass filter is a mass spectrometer in which the ions are separated by passing them through an oscillating quadrupolar field. Only ions of a particular mass/charge ratio can traverse the system, the others executing unstable trajectories and thus not being transmitted. In principle at least it can distinguish between any chemically-different species present and even between ions of the same nominal mass (such as O^+ and CH_4^+) provided the resolution is adequate. The main drawback is that its efficiency is lower by several orders of magnitude than that of the surface ionization detector. One should perhaps not be too pessimistic about the future of such detectors: efficiencies of ion sources are continually being improved and, already, the source of a commercially available quadrupole mass spectrometer is quoted as having an efficiency of 10^{-3} in ionizing a beam of nitrogen molecules.

Crossed-beam apparatus which make use of mass spectrometer detection, nozzle sources, and wide angle variations have recently come to be termed 'supermachines'.

Crossed-beam experiments

The ideal kinetic experiment would involve measuring a cross-section for every conceivable energy distribution, including both internal and external modes, and every orientation of both reactant and product species. The crossed molecular beam technique cannot yet achieve this ambitious goal, and indeed it is doubtful whether such experiments would be particularly desirable, bearing in mind the enormous amount of numerical data which would be generated. Present-day goals are restricted to the specific aspects of reaction dynamics listed below.

The variation of reaction probability with energy

Most elementary theories of reaction kinetics treat the energy requirement as a 'stop-go' criterion i.e. if the collision pair contains less than a certain energy minimum reaction cannot occur, whilst if it exceeds this value then reaction is inevitable. Although such a description is adequate for dealing with an overall rate constant, it is certainly an over-simplification in regard to a single encounter. As a corollary, the experimental activation energy obtained from the temperature coefficient of the bulk rate constant does not necessarily correspond to the true minimum, or *threshold*, energy below which reaction can never occur.

In principle the crossed molecular beam method is ideally suited to measuring the energy dependence of reaction since the relative kinetic energy of the reactants is directly controlled by the velocity selector. However, the relatively low energy of effusive sources means that so far the technique has been restricted to systems with extremely low threshold energies. Such measurements have produced threshold energies varying from zero for $K + DBr \rightarrow KBr + D$ to 2.4 kJ mol^{-1} for $K + HCl \rightarrow KCl + H$.

The variation of reaction probability with orientation

One factor believed to influence reaction probability is the orientation of the collision partners at impact; this corresponds to the *steric factor* of simple collision theory. It is possible to orient polar molecules in an inhomogeneous electric field; using this technique it has been shown that the rate of the reaction

$$K + CH_3I \rightarrow KI + CH_3$$

is almost doubled when the potassium atoms strike the iodine end of the methyl iodide molecule. Although in qualitative terms such a finding is hardly unexpected, this appears to be the only direct demonstration of the effect.

The dynamics of the collision

The significance of the dynamics of the collision may be appreciated by considering the following simple example in which a molecule A (depicted in Fig. 2.3 by an open circle) moving with a velocity v strikes a stationary molecule B of equal mass (depicted by the filled circle). There are three possible outcomes of an elastic collision in which translational energy is conserved and no chemical reaction occurs.

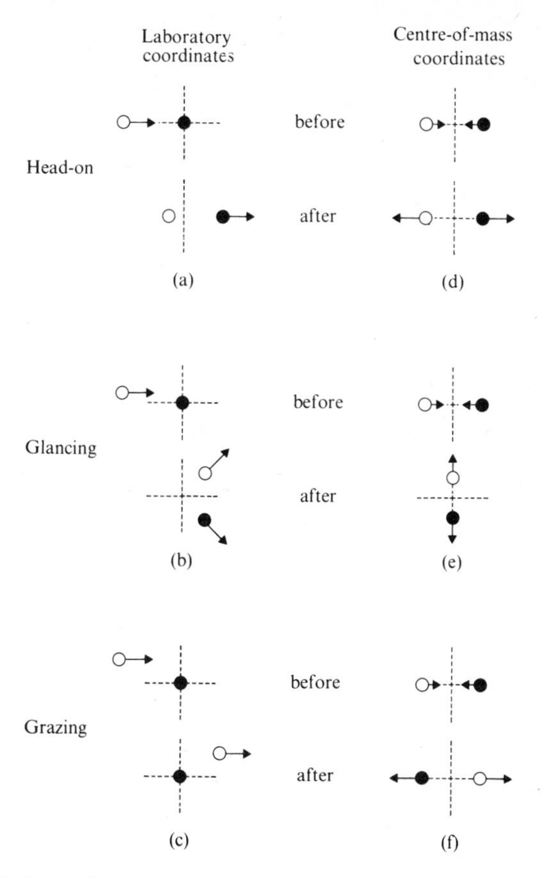

Fig. 2.3. Dynamics of elastic collisions in laboratory and in centre-of-mass coordinates.

(1) if the collision is head-on, molecule A is brought to a standstill and molecule B takes up the velocity v.
(2) If the collision is glancing, both particles move away at an angle to the line of impact.
(3) If the collision is grazing, such that no significant interaction occurs, molecule A passes B and retains its original velocity v.

These three situations are illustrated in Fig. 2.3a, b, c. Because the system originally possessed a total linear momentum mv, which must be retained, all resulting species appear in a forward direction and cluster closer to the initial direction as the velocity increases. Consideration of scattering in the laboratory system of co-ordinates thus provides very limited insight into the collision dynamics.

This difficulty can be overcome by depicting the collision in *centre-of-mass co-ordinates*, that is with respect to a co-ordinate system whose origin is sited at the centre-of-mass. The diagrams 2.3d, e, and f show how the three situations 2.3a, b, and c appear in the new representation. The concept is readily extended to a typical reactive situation by assuming that an atom or fragment from molecule B is transferred to molecule A. The subsequent motions of A and B are then characteristic of product species. (By means of a *Newton diagram*, whose construction need not concern us here, it is possible to convert laboratory measurements of scattering intensity with angle into polar diagrams which illustrate the results of the interaction in centre-of-mass co-ordinates.)

The angular distribution of products as a function of scattering angle in the laboratory system is shown in Fig. 2.4: these results are typical

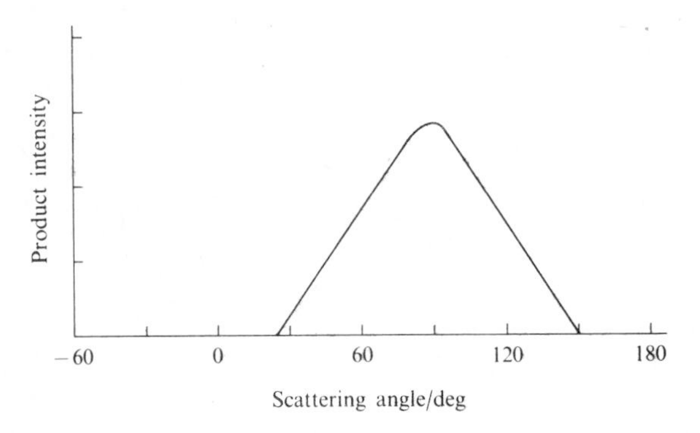

Fig. 2.4. Angular distribution of products in Type I reactions.

of KI formed by the reaction of K + CH_3I. Transformation to the polar diagram in Fig. 2.5 demonstrates that reactive scattering occurs in the backward hemisphere. This implies that reactive events took place by near head-on collisions. Such situations are termed *Type I* or *rebound* mechanisms; typical cross-sections lie in the range 0.05 - 0.75 nm^2 (5 - 75 $Å^2$).

In complete contrast, the reaction of K + Br_2 to produce KBr and other similar alkali metal plus halogen reactions yield polar diagrams (Fig. 2.6) characteristic of forward scattering. Reaction has thus taken place at large intermolecular separations where grazing collisions occur. These *Type II* reactions are strikingly termed *spectator-stripping* or

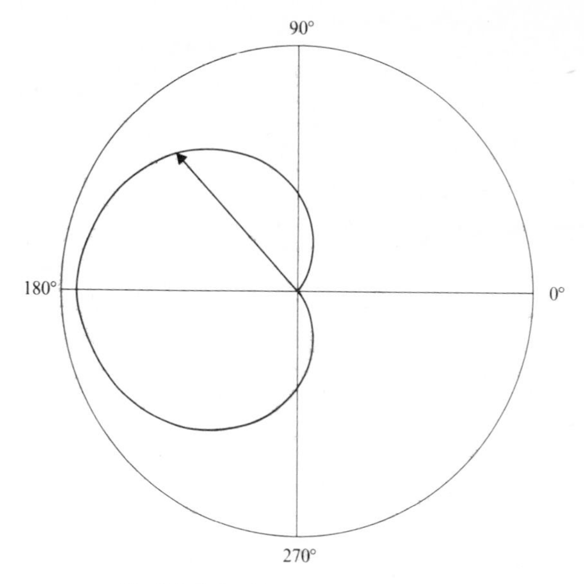

Fig. 2.5. Polar diagram for Type I reactions.

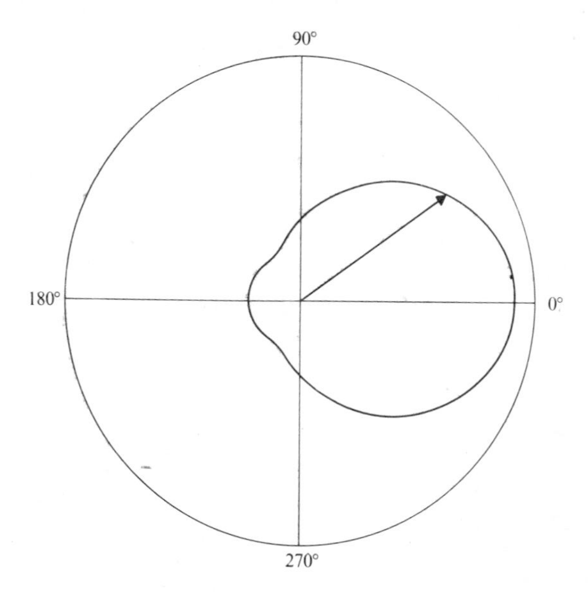

Fig. 2.6. Polar diagram for Type II reactions.

harpooning mechanisms and possess cross-sections greater than 2 nm^2 (200 Å^2). The notion involved in harpooning is that an electron (the harpoon) from one species transfers to the other species at relatively long range. The subsequent trajectories are then modified by Coulombic interaction (the line attached to the harpoon) of the charged species thus created.†

An intermediate situation in which both forward and backward scattering are observed, can be attributed to a rather different cause (see following section).

The duration of the collision

Many theories postulate the occurrence of a 'sticky' collision with the formation of a long-lived collision complex. Such a complex may be said to exist when the collision partners remain in close proximity for a time comparable to or greater than the period of rotation. In the discussion above it was shown that, for simple collisions, the species retain a memory of the original direction and the products tend to appear in either the forward hemisphere or the backward hemisphere with respect to a centre-of-mass co-ordinate system. However if the lifetime is long compared to that required for rotation the system 'forgets' the

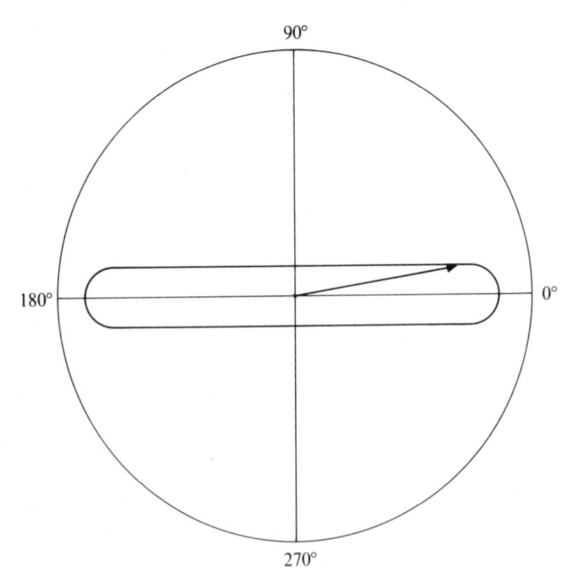

Fig. 2.7. Polar diagram for Type III reactions.

† See Levine and Bernstein, *Loc. cit.*

original orientation and the products depart with an equal probability in any direction. Although most of the systems so far studied do not show this behaviour, such symmetry has been observed in the reactions of alkali metals with alkali halides. For example, the reaction of caesium with rubidium chloride leads to the polar diagram shown in Fig. 2.7; such behaviour is termed a *Type III mechanism*. Although it is not yet known how general is such complex formation, we can at least be assured that it is not unreasonable to postulate the existence of such long-lived species.

The distribution of product energy

Most of the reactions studied in crossed-beam experiments are exothermic and it is of interest to determine how the energy released is distributed between the translational, rotational, and vibrational modes available.

Translational energy. The most direct method of determining the translational energy of the molecules employs a *velocity analyzer* identical to the velocity selector used to generate a monoenergetic primary beam.

Alternatively the translational energy may be obtained from the angular distribution of the scattered product molecules. This may be appreciated from the collision depicted in Fig. 2.8. A typical collision between two bodies of equal mass, one stationary and the other travelling with a velocity v, has resulted in both separating with a velocity $v/2$ at an

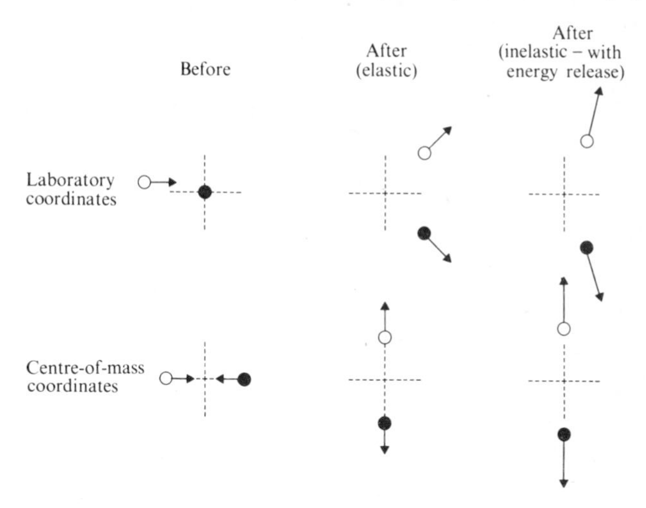

Fig. 2.8. Effect of energy release on collision dynamics.

angle of 45° to the original direction of motion. The forward and perpendicular components of linear momentum have remained unchanged at mv and zero respectively and the energy of $\frac{1}{2}mv^2$ is also unaffected. If energy is now released into the collision by reaction, the only way in which the components of momentum are conserved is by an increase in the scattering angle. Thus, if the particles leave at an angle of 60° with a velocity v, the two components of momentum are still the same but the energy has doubled. Although this is a rather special case, and the actual analysis of the experimental data is complex, it does illustrate how the angular dependence of product intensity can provide a measure of the energy released in translation.

Rotational energy. Provided the product molecules are polar, their rotational energy can be determined by passing them through an inhomogeneous electric field. The principle here is that the effective dipole moment of a polar species is decreased by rotational averaging so that the molecules least deflected are those with the highest rotational energy.

Vibrational energy. If the translational and rotational energies of the molecules can be measured, the vibrational energy is given directly by the conservation equations. In some cases, vibrational energies can be obtained by spectroscopic methods, for example by allowing the products to undergo second scattering in which the vibrational energy is transformed to electronic excitation and hence to light emission. The latter technique was employed first in the sodium flame experiments discussed in Chapter 7.

The determination of total reaction cross-sections and potential energy surfaces

The total reaction cross-section can be related to the conventional kinetic rate constant (see Fig. 1.3) and is clearly the next parameter which one would like to obtain from crossed-beam experiments. Such measurements prove very difficult because, in contrast to the features discussed above, they must be made on an absolute rather than a relative basis. This means, for example, integrating over all possible scattering angles. The exercise has therefore been carried out only in a small number of cases.

Measurements of elastic and inelastic scattering are used to obtain intermolecular potentials; for reasons which need not concern us here, 'fine structure' which appears in the angular distributions provides quite detailed information about these potential functions.

Reactive scattering measurements should provide similar information about potential energy surfaces for reacting systems. Although such a goal is beyond present techniques, calculations based on empirical surfaces provide numerical values for the quantities described above, namely the threshold energy for reaction, the effect of orientation, the partition of energy among the various modes, and the total reaction cross-section. Thus it is possible, by comparison with the experimental results, to establish qualitatively the features required of such surfaces in the critical region governing the reactive event.

In the reaction of K with CH_3I, which has been extensively studied, it is possible to determine the spacing and position of the contours which determine the actual dynamics of the reaction. The relationship between the potential energy surface and the dynamics of the collision is discussed again in the following chapter where the results of experiments based on a different technique are described.

3. Infrared chemiluminescence

AT the end of the previous chapter, it was shown how information on collision dynamics is able to reveal the detailed structure of potential energy surfaces. This type of study has been considerably extended by a rather different type of experiment based on the radiation emitted by the products of reaction. Such experiments yield very specific information on the partitioning of the reaction energy among the various modes while at the same time employing appreciably simpler equipment than required for crossed-beam work. Infrared chemiluminescence studies in association with computer simulations have produced quite exciting data on the processes which occur during a chemical reaction. The success can be attributed directly to this 'two-pronged' attack on the problem. The results of the computer simulations are, for fairly obvious reasons, somewhat in advance of the experimental work which means that some of the theoretical predictions have not yet been fully tested by experiment.

The experimental approach

The principle underlying these techniques is that molecules with permanent dipole moments which are formed in vibrationally-excited states emit infrared radiation characteristic of their particular vibration-rotation level. The critical feature of each technique is the method employed to measure the population distribution of the product molecules immediately after their formation, and prior to any significant relaxation to the ground state.

In the initial studies, the reagents were mixed as rapidly as possible and then swept past an observation window situated close to the mixing point. This procedure was subsequently refined by spacing a series of observation windows along the flow tube downstream of the mixing point. Such a modification provides a considerable improvement because instead of simply neglecting the effect of relaxation, it becomes possible to measure its rate and hence extrapolate back to the situation which existed immediately after mixing. This is termed the *measured relaxation* method.

An alternative way of tackling the problem which was developed later, is to mix molecular beams of the reagents, rather than bulk flows, in the centre of a large evacuated chamber. This chamber is cooled to liquid nitrogen temperature (77 K) or below, so that the product molecules which arrive there are either removed cryogenically or reside there for sufficient time to release their excess energy and revert to the ground state. By this means, the excited product molecules undergo a negligible number of secondary collisions and the emission truly corresponds to the initial situation. The success of this technique derives from the molecular beam principle, although well-collimated beams are not required and the intensity problems are less severe. The effectiveness of this *arrested relaxation* technique has been demonstrated by comparison with measured relaxation measurements and, more strikingly, by observing that the initial rotational, as well as vibrational, distributions are 'frozen'.

A recent development makes use of the laser action which can arise as a result of the non-equilibrium population distribution. Laser techniques for measuring the relative rates of populating different vibrational levels rely on two important principles. The first is that, provided collisional deactivation is negligible or can be allowed for, the ratio between the populations of two adjacent vibrational levels N_v/N_{v-1} is equal to the ratio of the rate constants for formation of these states, k_v/k_{v-1}. The second is that the relative intensities of various vibration-rotation transitions depend on the distribution of energy among the rotational levels and hence on the 'rotational' temperature.

In the earliest work, which involved an HF product, the exothermicity of the reaction caused the temperature to increase with time and hence a sequence of laser emissions occurred, arising from different vibration-rotation transitions. By identifying the nature of these transitions and calculating the heat evolved and therefore the change in rotational temperature between the onset of the different transitions, it was possible to estimate N_v/N_{v-1} within quite narrow limits (e.g. 0.83 - 1.11).

The disadvantages of this procedure are that it requires a knowledge of the rotational temperature as it changes with time during the experiment and that it succeeds only in establishing bounds for N_v/N_{v-1}. An improvement, known as the *equal gain* method, is to adjust the rotational temperature by external heating until two laser transitions are initiated simultaneously and with equal intensity. This yields the rate constant ratio k_v/k_{v-1} with high precision and is therefore likely to be used more frequently in the future.

The majority of infrared chemiluminescence studies have involved hydrogen halide systems. In the reaction between hydrogen atoms and chlorine molecules, which has been extensively studied,

$$H + Cl_2 \rightarrow HCl^* + Cl$$

the exothermicity is 203.2 kJ mol^{-1} (48.6 kcal mol^{-1}). 40 per cent of this exothermicity appears in vibrational excitation, the relative values of the rate constants for population of the various levels being: $k(v=1)$ 0.31; $k(v=2)$ 0.61; $k(v=3)$ 1.00; $k(v=4)$ 0.22; $k(v=5)$ 0.03; $k(v=6)$ 0.003 (These figures were obtained by the measured relaxation method: other methods give qualitative agreement but the numerical values differ somewhat.) In general, for reactions involving hydrogen or deuterium atoms, e.g.

$$D + Cl_2 \rightarrow DCl^* + Cl$$

$$H + Br_2 \rightarrow HBr^* + Br,$$

about 40 - 50 per cent of the energy appears in vibration. For reactions of halogen atoms, e.g.

$$F + H_2 \rightarrow HF^* + H$$

$$F + D_2 \rightarrow DF^* + D$$

$$Cl + HI \rightarrow HCl^* + I$$

$$Cl + DI \rightarrow DCl^* + I,$$

the fraction of the exothermicity which appears in vibration is closer to 70 per cent. In all cases, the remaining excitation appears in translation rather than rotation. The replacement of hydrogen by deuterium markedly affects the absolute rate constants but the energy distribution is largely unaltered.

The theoretical approach

The use of potential energy surfaces to describe the progress of a chemical reaction was referred to briefly in the previous chapter and is discussed in greater detail in a companion volume (*Reaction Kinetics* by M. J. Pilling, O.C.S. 22). If the potential energy surface for a reaction, that is the energy for any geometrical configuration of nuclei, is known the probability of reaction for a particular encounter can be evaluated. In practice, quantum-mechanical calculations provide surfaces accurate to about 10 kJ mol^{-1} (equivalent to 10 per cent in the activation energy, 1 per cent in the binding energy, or 0.05 per cent in the total energy) only for the H_3^+ and H_3 systems. The approach employed is therefore to construct semi-empirical surfaces, which preferably have some theoretical foundation, and then compare predictions based on such surfaces with experimental data. In order to make these predictions it is necessary to start with a representative point moving on the surface, corresponding to one particular set of initial conditions, and then to determine its trajectory by carrying out a stepwise integration based on the appropriate, usually classical, equations of motion. This procedure should really be repeated with every possible representative point associated with the various possible orientations, velocities, impact parameters, and internal energy states, but the amount of computer time required would be quite overwhelming. This difficulty is avoided by the *Monte Carlo* technique; a series of representative points are selected at random, subject to the statistical distribution required by the temperature. Such computations turn out to converge quite rapidly.

The first major feature illustrated by such calculations and confirmed by experiment has already been mentioned; if the 'downhill' part of the reaction coordinate lies along the approach coordinate r_{AB} the surface is termed *attractive*, and the exothermicity appears as vibrational excitation of the new bond (Fig. 3.1a). On a *repulsive* surface (Fig. 3.1b), however, the energy appears in translational motions of the products.

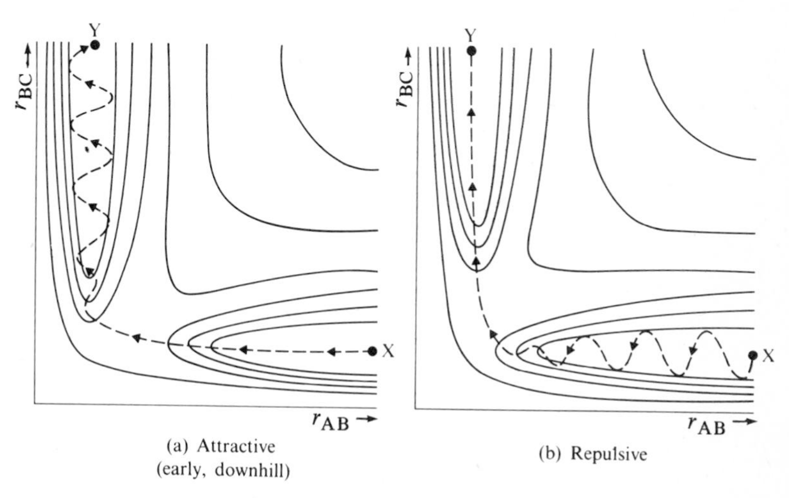

Fig. 3.1. Attractive and repulsive potential energy surfaces.

One refinement of this concept is to note that, along part of the path, both r_{AB} and r_{BC} are changing so that the path is 'curved' and not rectangular. The curved portion is responsible for *mixed energy release* in which both translational and vibrational excitation might be expected. Another feature which must be taken into account is the combination of masses involved. Where the masses are roughly comparable, mixed energy release turns out to favour internal excitation of the products, but if the attacking atom is light (e.g. in $H + Cl_2$) more of the energy is likely to appear in translation. This is termed the *light-atom anomaly*.

Detailed analysis of the trajectories on various surfaces shows that even in the absence of deformities, i.e. dents, hollows, etc., the systems frequently undergo *complex* or *indirect* encounters in which the species begin to separate and then return to undergo one or more secondary encounters. Such events may lead to 'clutching' or 'clouting' depending on whether the products are pulled together or thrown apart.

4. Flash photolysis

ONE of the best-known methods for the study of fast reactions is the technique of flash photolysis, reported first by Norrish and Porter in 1949 and for which they shared a Nobel prize in 1967. This technique falls in the category of large-perturbation, real-time methods described in the introductory chapter.

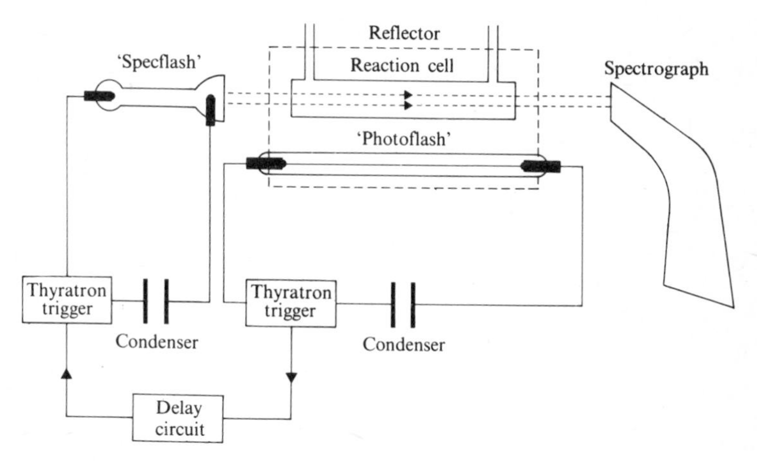

Fig. 4.1. Flash photolysis apparatus.

The basic experimental arrangement, illustrated in Fig. 4.1, has changed little since its inception. The reaction vessel comprises a quartz cylinder up to 1 m in length with plane windows at each end. One or more *photoflash* tubes are situated parallel to it and the complete assembly is surrounded by a suitable reflector, usually coated with magnesium oxide. If experiments have to be conducted at a variety of temperatures, the apparatus is completely immersed in a thermostatted bath.

The photoflash lamp consists of a long quartz tube fitted at each end with metal electrodes, commonly of molybdenum, and containing a low pressure of inert gas. The light flash is generated by discharging a large condenser, previously charged to a high potential, across the two electrodes. Although the particular arrangement will depend on the application, a typical installation might employ two 10 μF condensers charged to 10 kV giving a flash energy of 1000 J. The conversion to useful radiant energy in the ultraviolet is quite efficient and produces about 10^{20} quanta corresponding to about one-tenth of the total energy dissipated. Provided the inductance of the circuit is kept low and the conductivity of the leads is high, the duration of the flash (to half the maximum intensity) is about 20 μs. In general the duration depends on the energy dissipated, varying from nanoseconds for microjoule flashes to milliseconds for flashes of a hundred thousand joules, although the dependence on capacitance is greater than on the applied potential. Since scattered light from the flash interferes with spectroscopic measurements, the time resolution of the technique depends on the duration of the flash, and it is usually necessary to effect a compromise between the conflicting demands of maximum conversion and optimum time response. In the earlier studies, the inert gas pressure was adjusted to 'hold off' the applied potential and the lamp was triggered by applying a high tension pulse to a third electrode. Nowadays it is more common to provide an external gap in series with the lamp, such as the thyratron† shown in the sketch; only a small fraction of the available energy is dissipated in such a device. The typical flash energy quoted is sufficient to produce a very high extent of conversion in a low pressure gas. Although the technique is necessarily restricted to molecules with suitable absorption characteristics, the light output is essentially a continuum down to the quartz cut-off so that the range of reactions which can be investigated is clearly very large.

The subsequent events must be followed in real time and hence the techniques available are rather limited. Infrared absorption and mass spectrometry have been employed with some success but by far the most popular method, developed concurrently with flash photolysis, is *kinetic spectroscopy*. Ideally, of course, one would like to scan the absorption spectrum repeatedly after the flash but this is ruled out by signal/noise problems and it is necessary either to monitor a narrow spectral band continuously or to obtain a single spectrum at a predetermined interval after the flash. The latter has proved the most profitable, at least until the reaction mechanism has been fully analyzed, the experiment being repeated with different time delays until the complete pattern of events has been resolved.

The absorption spectrum is obtained by utilizing a second flash lamp, sometimes termed the *specflash*, situated at one end of the reaction vessel. This lamp is simply a smaller version of the photoflash, frequently with the discharge contained in a narrow capillary. By using the same

† A thyratron is basically a gas-filled thermionic valve. A suitably-biassed grid enables it to withstand a high potential difference but, once triggered, the grid loses control and the valve can carry a heavy current.

power supply but with a 4 μF condenser, a flash energy of 200 J coupled with a duration of 2 - 3 μs is obtained. This provides adequate light for the spectrograph, situated at the opposite end of the reaction chamber, but insufficient to interfere with the photochemistry.

It should be noted that the light energy absorbed by the system, although initially leading to electronic excitation, must eventually be degraded into heat. In pure gases, this may lead to temperature increases of several hundred degrees. Such an *adiabatic* system has been of value in some studies but in most cases *isothermal* conditions are achieved by diluting the reactants with a large excess, perhaps 500-fold, of inert gas. The temperature rises in liquid systems are much smaller so that in them reactions normally occur isothermally.

The experimental arrangement described here permits the study of first order reaction rate constants up to 10^6 s^{-1} and second order rate constants up to 10^{11} l mol^{-1} s^{-1}.

Identification of transient intermediates

The measurement of the rates of fast reactions and the detection and identification of transient species are closely interrelated problems which are not really separable but it is fair to say that, in contrast to most other techniques described in this book, flash photolysis has proved particularly successful in the latter role. This is not in any way to deny the valuable kinetic measurements which have been made. However, the difficulty of tracing the route by which the initial electronic excitation is degraded into translational, or temperature, energy, of achieving controlled, uniform temperatures, and of eliminating surface effects means that other techniques can frequently produce kinetic data of comparable precision or reliability. On the other hand, no other approach can come near to matching the success which flash photolysis has had in identifying transient intermediates. Table 4.1 lists some of the most significant species first identified by flash photolysis. The particular

TABLE 4.1

Examples of intermediates identified by flash photolysis

CS	ClO	C_3	HCCl
CN	PH	CH_2	HSiCl
CH	NH_2	CHO	HSiBr
NH	N_3	CF_2	BO_2
S_2	NCO	HS_2	$PhCH_2$
SO	NCN	HNO	PhNH
SH	NCS	HCF	PhO

value of the technique stems from the ability to use intense flashes, of duration much shorter than the lifetime of the species under examination, and to provide extremely long optical path lengths for spectroscopic detection. The positive identification of the gaseous radicals is usually accomplished by careful analysis of the complex pattern of lines produced, whilst in the case of aromatic species, whose spectra cannot be resolved, it is necessary to rely on comparisons between a series of related radicals.

Production of vibrationally-excited species

One of the most significant discoveries from gas-phase flash photolysis has been the production of molecular species in vibrationally-excited states. The photolysis of nitrogen dioxide showed the presence of vibrationally-excited oxygen with v'' up to 11, while ClO_2 yielded O_2 ($v'' = 8$) and O_3 gave v'' up to 17. In each case this phenomenon can be attributed to subsequent reactions of oxygen atoms with the parent molecule.

Chlorine dioxide: $ClO_2 + h\nu \rightarrow ClO + O$

$$O + ClO_2 \rightarrow O_2{}^* + ClO + 255\ \text{kJ mol}^{-1}\ (61\ \text{kcal mol}^{-1})$$

Nitrogen dioxide: $NO_2 + h\nu \rightarrow NO + O$

$$O + NO_2 \rightarrow O_2{}^* + NO + 192\ \text{kJ mol}^{-1}\ (46\ \text{kcal mol}^{-1})$$

Ozone $O_3 + h\nu \rightarrow O_2 + O$

$$O + O_3 \rightarrow O_2{}^* + O_2 + 577\ \text{kJ mol}^{-1}\ (138\ \text{kcal mol}^{-1})$$

A further series of studies showed that oxygen atoms react with simple hydrides to produce OH radicals in vibrationally-excited states

$$O + H_2 \rightarrow OH^* + H$$

$$O + HCl \rightarrow OH^* + Cl$$

$$O + NH_3 \rightarrow OH^* + NH_2$$

$$O + CH_4 \rightarrow OH^* + CH_3.$$

Although these reactions all involve oxygen atoms it has been found that hydrogen atoms, halogen atoms, and sodium atoms undergo analogous reactions. It has therefore been possible to make the generalization that, in an exothermic reaction of the type

$$A + BCD \rightarrow AB + CD,$$

a large fraction of the energy liberated appears as vibrational excitation of the newly-formed (AB) bond. The production of these vibrationally-excited species also permits investigation of the subsequent energy-transfer processes.

Recombination of atoms

Another example in which flash photolysis has proved invaluable is in the study of atom recombination reactions, notably that of iodine atoms. The flash serves to dissociate halogen molecules into atoms with consequent reduction in absorption, the subsequent reappearance of the diatomic species providing a measure of the rate of recombination. The studies revealed that the recombination is third order, as had been predicted theoretically, that there are considerable differences between the efficiencies of various molecules acting as 'third bodies', and finally the surprising result that the measured rate constants display *negative* temperature coefficients. The eventual conclusion was that an intermediate complex must be formed between one of the atoms and a 'third body'. As the stability of the complex decreases with increasing temperature, the overall rate falls with a (negative) temperature coefficient related directly to the binding energy of the complex. (When expressed in Arrhenius form this temperature coefficient takes the form of a 'negative activation energy'.) The mechanism is then

$$\mathrm{I + M \rightleftharpoons IM}$$

$$\mathrm{IM + I \rightarrow I_2 + M.}$$

In some cases complex formation is due to charge transfer and characteristic spectra have indeed been observed. The most efficient *chaperon* discovered was nitric oxide and the spectrum of the molecule NOI was observed.

Measurement of elementary reaction rates

A major problem with optical absorption methods is that, when the concentration of absorber is low, the difference between two relatively large light intensities has to be measured and the precision is therefore low. By illuminating with a source of characteristic resonance radiation, the species of interest can be made to fluoresce in all directions. The light output at right angles to the incident beam is then directly proportional to concentration. Repeated experiments, combined with pulse-counting techniques, provide accurate decay curves at very low atom concentrations and hence high reactant excess. This *resonance-fluorescence* technique yields measurements of elementary reaction rates, e.g. O atoms with olefins, with a remarkable accuracy and reliability not previously associated with flash photolysis methods.

Primary photolysis processes

Another major application of flash photolysis has been in the study of primary processes in photoexcitation; this application also serves to illustrate the use of flash photolysis in the realm of solution kinetics.

It was demonstrated in the early 1940s that the phosphorescence displayed by many solid organic molecules is due to light emission from a low-lying triplet state. This phosphorescence is not observed in the liquid or gas phase and the role, if any, of triplet states in such systems was therefore very much a matter for conjecture. Some ten years later, flash photolysis experiments demonstrated that identical triplet species

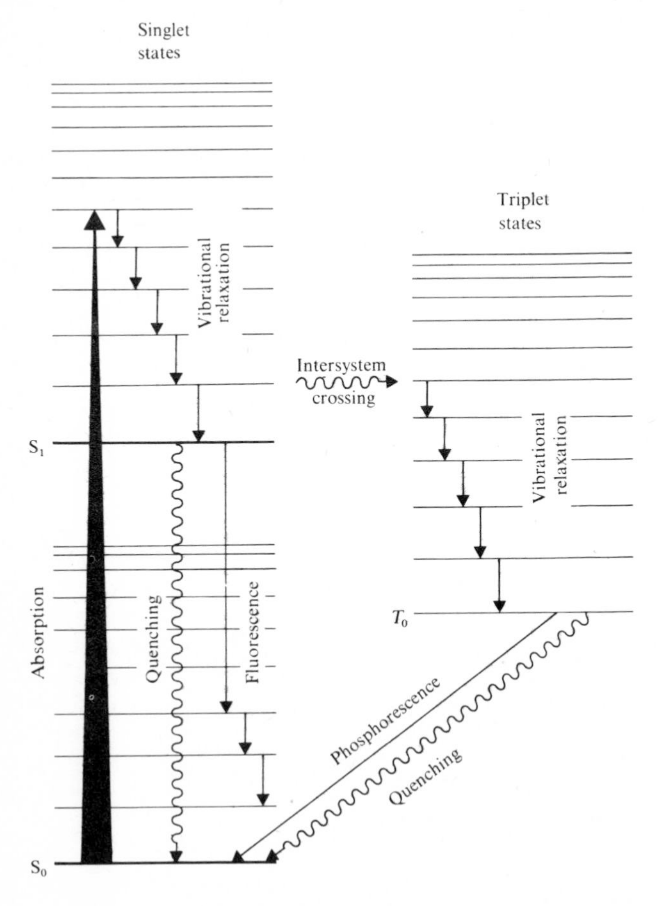

Fig. 4.2. Primary photolysis processes.

are formed in gases and in liquids, with lifetimes up to a millisecond provided oxygen is carefully excluded. These experiments were conducted with polynuclear aromatic hydrocarbons and their derivatives, although subsequent experiments have shown that virtually all photo-excited organic molecules show a similar pattern of behaviour.

The primary processes which occur when an organic molecule absorbs light† are indicated in the energy level diagram of Fig. 4.2. Initial excitation is to a series of singlet levels but, in solution, these rapidly convert by radiationless transitions to the lowest singlet state, S_1. This state may then undergo chemical reaction, fluoresce to the ground state S_0, or undergo intersystem crossing to the triplet state. The evidence so far available indicates that radiationless conversion from S_1 to S_0 is rarely important. In the same way the triplet state T_1 may undergo chemical reaction, radiate (phosphoresce) back to the ground state, or be *quenched* in some way. In order to gain an understanding of photochemical processes it is necessary to study the properties of the lowest singlet and triplet states. The latter possess lifetimes in the order of 10^{-6} to 10^{-3}s and can therefore be studied by flash photolysis. Singlet states with lifetimes down to 10^{-9}s can be studied only by nanosecond flash photolysis, which is described later; because of the longer lifetimes, many photochemical reactions of interest proceed via triplet intermediates. The actual spectrum of the triplet state is not particularly informative, but the decay characteristics, usually examined by monitoring a single wavelength, have yielded enormous amounts of information.

It became evident quite quickly that triplet species are very efficiently quenched by trace impurities, and the observation of phosphorescence in the solid state arises because impurities and triplet species are unable to diffuse together. More detailed investigation has revealed the following behaviour. In the gas-phase or at high concentrations in solution, second-order decay by *triplet-triplet annihilation* is the predominant process, leading to excited singlet states and delayed fluorescence. When this process is not important, first-order decay is observed and two distinct regions of behaviour occur. At low temperatures and high viscosity, conditions approaching the solid state, true radiative and radiationless conversion take place. At lower viscosities and higher temperatures an apparent first-order quenching occurs, controlled by the rate of diffusion through the solvent. Four distinct quenching mechanisms can be identified:

(1) Heavy atom quenching, due to an increase in spin-orbit interaction.

† See Atkins' *Quanta* (OCS 21) for an account of excitation of molecules, and their subsequent decay processes.

(2) Interaction with paramagnetic species to form complexes in which spin conversion is allowed.
(3) Formation of charge-transfer complexes.
(4) Transfer to a different molecule with lower-lying electronic energy levels.

The last of these may be recognized as an important step in any photoinduced reaction and forms the basis of the technique of *photosensitization.*

It should be appreciated that the excited singlet and triplet states have electronic configurations different from the ground state molecules and therefore possess a quite distinct chemistry, much of which can be investigated by flash photolysis.

Nanosecond flash photolysis

One of the most recent advances in the development of flash photolysis has been the extension of the time resolution from the microsecond range down to the nanosecond range. The source of the high energy, short duration flash is the Q-switched laser. Sources with adequate energy and short duration are readily available although the wavelength range is still relatively restricted.

The major experimental problem is to devise a specflash which produces a continuum light pulse of comparable duration at a predetermined interval, in the nanosecond range, after the laser pulse. An elegant way in which this has been achieved is shown in Fig. 4.3. The laser pulse is split into two parts, one of which passes into the reaction vessel and initiates the photolysis. The other beam is directed on to a moveable mirror and is reflected back to trigger the continuum source. One such source is a fluorescent solution (1, 1, 4, 4-tetraphenylbuta-1, 3-diene in cyclohexane) which has the effect of transforming the monochromatic laser pulse into continuous radiation in the visible region. An alternative consists of a simple spark gap which is maintained close to the breakdown potential so that ionization produced by the laser pulse triggers the discharge. Both sources give light pulses of comparable duration to the laser pulse. The time delay is very simply altered by moving the mirror towards, or away from, the apparatus: 15 cm travel alters the delay by a nanosecond.

The motivation for working in the nanosecond range follows from the earlier discussion of triplet states. The energy level diagram shows that triplet states arise after initial excitation of singlet states of much shorter lifetime. Only the triplets can be observed by microsecond flash photolysis and nanosecond flash photolysis provides an opportunity for carrying out comparable investigations on singlet states.

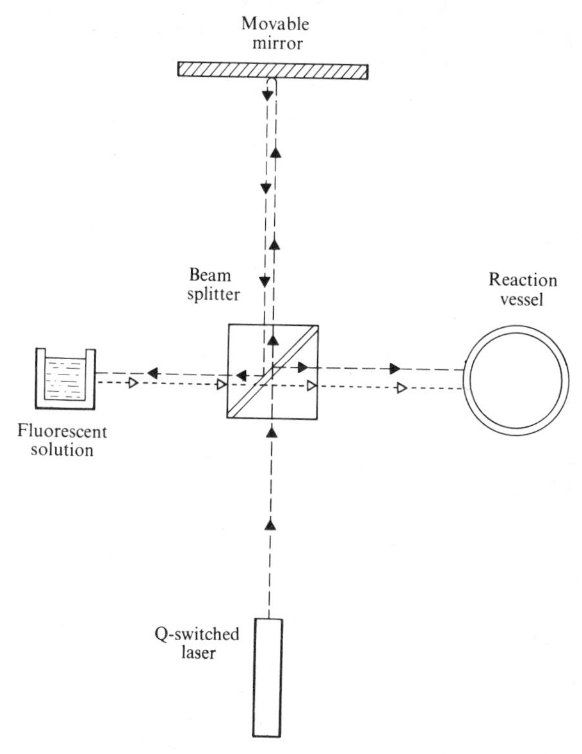

Fig. 4.3. Arrangement for nanosecond flash photolysis.

Pulse radiolysis

The basic flash photolysis technique has been extended in two directions: to shorter times, as in nanosecond flash photolysis, or to more energetic radiation, as in pulse radiolysis. The experimental principles underlying pulse radiolysis are essentially identical with those of flash photolysis and it is unnecessary to describe them in detail. It is useful, however, to draw attention to certain differences between the two types of investigation.

In photochemical experiments, typified by flash photolysis, the photons are selectively absorbed by molecules with appropriate spectral response and the formation of a molecule in an excited state is associated with the removal of a photon from the beam. In radiation chemistry, one is dealing with much more energetic species (typically > 50 eV) which may include α-particles, electrons, etc. as well as electromagnetic radiation in the form of γ- and X-rays. The energy is absorbed from such radiation in a non-selective manner and a single particle or photon will produce a variety of ions and excited states along its *track*. Because of the lack of selectivity this complex mixture of fragments will be characteristic of the solvent rather than of added reactants. Many of these fragment species possess sufficient energy to generate *spurs* of ionization and excitation separate from the tracks of the incident radiation.†

These factors have some influence on the design of the apparatus. To begin with the nature of the radiation has only a secondary influence on the subsequent reactions and one is therefore able to use the most convenient source which happens to be available.

Microwave linear electron accelerators

A magnetron or klystron is used to supply radio-frequency waves to a linear waveguide. Pulses of electrons are fed in at one end and are forced by the propagating waves to travel at progressively increasing velocity. Pulse durations of 0.5 to 5 μs with energies of one to fifteen MeV are normally used giving dose rates of 10^{18} eV cm^{-3}. Pulse durations down to 10 nanoseconds, with correspondingly higher energies, can also be achieved.

Van de Graaff electron accelerators

These accelerators can generate electron pulses with characteristics similar to those of linear accelerators, although the upper limits on current and energy are lower. They therefore prove less suitable if higher energy or shorter duration pulses are required.

Pulsed X-ray systems

X-ray systems operate at substantially lower energies and although short duration pulses can be achieved, the lower dose rates available limit the species which can be detected.

Because the solvent absorbs the radiation, and illumination as uniform as possible is needed if reliable kinetic studies are to be made (as opposed simply to detecting and identifying products), the cell volume must normally be kept small, usually not greater than about 10 ml. Much larger cells may be used for studies of gaseous systems.

Although detection of transient species has been achieved by electron spin resonance and electrical conductivity, optical detection similar to flash photolysis is the dominant method for following the subsequent processes. The only significant difference is that the small size of the

† For an account of radiation chemistry, see G. Hughes: *Radiation Chemistry* (OCS 6)

reaction cell makes multiple traverses essential to achieve the required amount of absorption.

Pulse radiolysis has been used to study organic and inorganic free radicals, molecular ions, and triplet excited states of aromatic molecules, but there is no doubt that its outstanding achievement has been in studies on the hydrated electron.

Reactions of the hydrated electron

The passage of ionizing radiation through water produces mainly positive H_2O^+ ions and secondary electrons. These species react with further water molecules to give H and OH radicals and the expected ions H_3O^+ and OH^-. The overall process for the production of H atoms may be represented by the equation

$$e^- + H_2O \rightarrow H + OH^-.$$

This reaction is endothermic in the gas phase so that it represents a feasible process only if the OH^- ion is hydrated. The time taken for such a process is not insignificant since it involves a re-orientation of the neighbouring molecules and in consequence the intermediate stage, wherein an electron is simply attached to a water molecule, has an appreciable lifetime. This *hydrated electron* e^-_{aq}, although of transient existence, turns out to be responsible for many of the reducing properties of irradiated water, rather than the hydrogen atom produced subsequently. From the experimentalist's point of view the intense blue colour characteristic of the solvated electron, which was well-known from studies on solutions of alkali metals in liquid ammonia, facilitates its study by pulse radiolysis and optical detection.

Although the mechanism of formation of the hydrated electron is considerably more complex than suggested above, about 10^{-9}s after the interaction with the ionizing particle the electrons prove to be fully thermalized and solvated. At the same time the solution contains comparable amounts of H_3O^+, OH, OH^-, H_2O_2, H, and H_2. These are listed in decreasing order of importance, with $[e^-_{aq}]$ being about equal to $[OH^-]$. The solvated electron is highly reactive and if its reactions with additives are to be investigated the effects of the other products of the ionizing radiation must be minimized. This can be achieved by adding methanol and sodium hydroxide, which together serve to remove H, OH, and H_3O^+, or by saturating with H_2 gas under alkaline conditions, which has a similar effect. The remaining reaction has a rate constant of 16 l mol^{-1} s^{-1} at 25°C. The slowest observable decay rate is governed by this reaction and reactions with added species can be studied provided the resulting decay rate is faster than this limiting rate.

TABLE 4.2

Some typical rate constants for reactions of e_{aq}^-

Reaction	Rate constant ($l\ mol^{-1}\ s^{-1}$)
$e_{aq}^- + OH \longrightarrow OH^-$	3×10^{10}
$e_{aq}^- + O_2 \longrightarrow O_2^-$	2×10^{10}
$e_{aq}^- + Cd_2 \longrightarrow Cd^+$	5×10^{10}
$e_{aq}^- + N_2O \longrightarrow N_2 + O^-$	9×10^9
$e_{aq}^- + H_2O \longrightarrow H + OH^-$	1.4×10^{10}
$e_{aq}^- + NH_4^+ \longrightarrow H + NH_3$	$\sim 10^6$

The rates of reaction of e_{aq}^- have been measured for a very wide range of compounds, some of them being listed in Table 4.2. It is interesting to note that many of them have rates in the range 10^9 - 10^{10} $l\ mol^{-1}\ s^{-1}$. An interesting question arises over the extent to which the rates are controlled simply by the speed at which the species can diffuse together. The answer is not entirely clearcut but the rates are sufficiently close to the diffusion-controlled limit to indicate that diffusion must be a significant, if not the sole, factor.

For further information about the hydrated electron see G. Hughes: *Radiation chemistry* (OCS 6) and G. Pass: *Ions in solution* (3): *inorganic properties* (OCS 7).

5. Shock waves

THE shock wave technique falls in the large perturbation, 'real-time'† category: the system is subjected to a very severe disturbance, in this instance of the translational energy, or temperature, and its subsequent relaxation to a new equilibrium position is monitored. The technique can be applied to liquids and solids but, in these materials, the equations of state dictate that the pressure, rather than the temperature, is strongly perturbed. As reaction rates are more sensitive to temperature, the major applications have been to gaseous systems.

A shock wave is a step transition in the properties of a fluid medium (Fig. 5.1) and propagates without change of shape: all the properties of

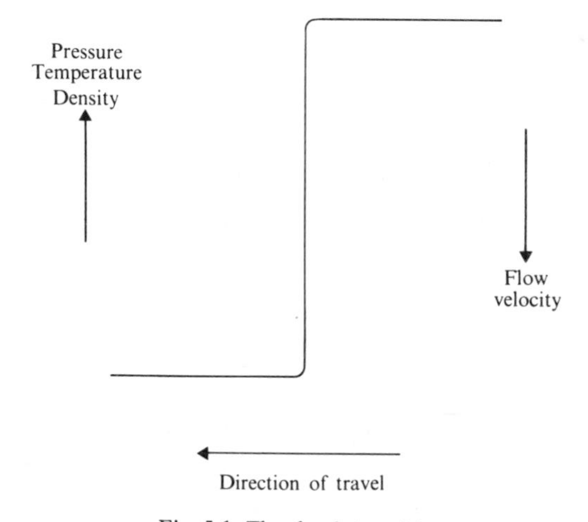

Fig. 5.1. The shock transition.

the 'fluid'—velocity, temperature, pressure, and density—are related to each other by conservation equations, and hence all change simultaneously. Although any fluid medium is really composed of discrete particles, the transition is as close to being a true discontinuity as the corpuscular nature of matter permits. Thus the 'thickness' of the shock front is equivalent to two or three mean free paths in the undisturbed

† If the measurements are made on the moving gas behind the shock front, laboratory time is not equal to real, or *particle*, time but may be converted to it by multiplying by a numerical factor of about four.

gas ahead (i.e. 10^{-5} cm at atmospheric pressure). Because the wave propagates at supersonic velocity, the time required for the transition is typically 10^{-9}s at atmospheric pressure. A feature of special significance for kinetic studies is that the transition occurs homogeneously in the gas phase and energy is transferred uniquely into translational motions of the molecules. The other energy modes are unexcited immediately behind the shock front and subsequently relax to their new equilibrium populations roughly in sequence of increasing energy i.e. rotation, vibration, dissociation and chemical reaction, electronic excitation, and ionization.

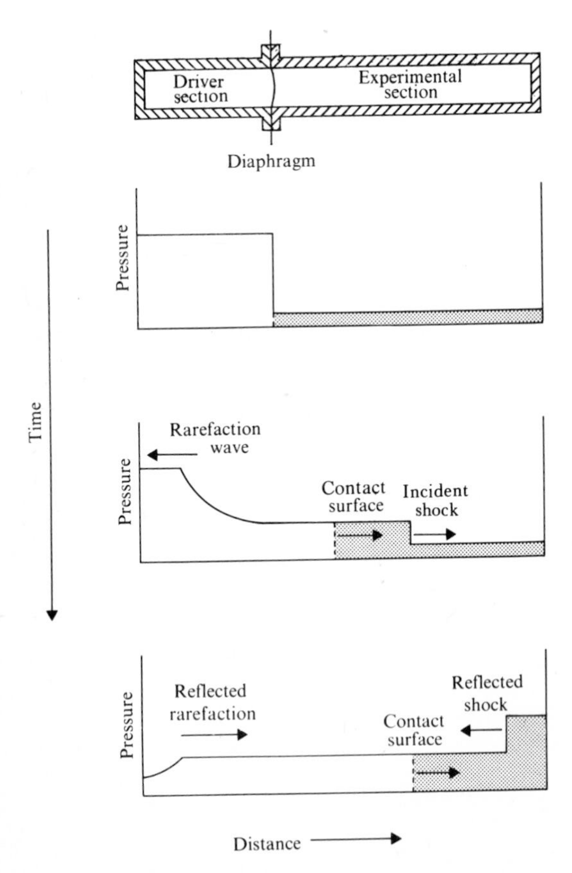

Fig. 5.2. Pressure profiles in conventional shock tube.

The shock tube

Shock waves can be generated in a number of ways, but the only method of real importance for the study of fast reactions involves the *shock tube*, of which there are three basic configurations in common use. The conventional shock tube illustrated schematically in Fig. 5.2 consists of a long tube of uniform cross-section divided into two parts by a thin diaphragm. Before the commencement of the experiment the two chambers are charged with gas, with the pressure on one side of the diaphragm sufficiently great that it is almost at bursting point. The actual rupture is effected either by raising the pressure on the high-pressure, or *driver*, side even further, or by piercing the diaphragm with a sharp needle. The high pressure gas then acts as an accelerating piston and drives a shock wave ahead into the low pressure, or *experimental*, gas. Subsequent pressure profiles are illustrated in the diagram. The shock wave does not form instantaneously but arises from the coalescence of pressure, or sound, pulses of low amplitude. Since sound waves lead to adiabatic heating the characteristic velocity is raised by each pulse and successive pulses therefore catch up their precursors.

An analysis of the shock wave phenomenon provides an interesting application of the conservation equations. Properties are defined with respect to an observer, or co-ordinate system, situated at the moving shock front. The undisturbed gas enters the front at the speed of the shock u_1 and leaves at a lower velocity u_2. If the properties ahead are denoted by subscript 1 and those behind by subscript 2, the equations for conservation of mass, momentum, and energy may be written respectively as:

$$\rho_1 u_1 = \rho_2 u_2 \text{; conservation of mass (continuity)} \quad (5.1)$$

$$p_1 + \rho_1 u_1^2 = p_2 + \rho_2 u_2^2 \text{; conservation of momentum} \quad (5.2)$$

$$e_1 + p_1/\rho_1 + \tfrac{1}{2}u_1^2 = e_2 + p_2/\rho_2 + \tfrac{1}{2}u_2^2 \text{; conservation of energy.} \quad (5.3)$$

In addition the properties of a gas are governed by two equations of state which define the pressure and the internal energy in terms of density and temperature i.e.

$$p = f_1(\rho, T) \quad (5.4)$$

$$e = f_2(\rho, T). \quad (5.5)$$

For an ideal gas that does not experience a mole number change, the first of these is simply $pu = nRT$ or $p = \rho RT$ and the second may be written $e = \int_0^T C_u \cdot \mathrm{d}T$. All properties denoted by the subscript 1, apart from the velocity u_1, may be determined prior to diaphragm rupture;

a count of the remainder shows that these five equations contain six unknowns so that the measurement of one parameter, usually the velocity, completely determines the rest.

The initial conditions on the two sides of the diaphragm roughly determine the subsequent history of the shock but cannot be used for accurate predictions because of indeterminate energy losses associated with the diaphragm rupture. For accurate kinetic measurements, it is normal to dilute the reactants with an inert gas so that the subsequent temperature changes due to reaction are minimized.

A typical shock tube has an internal diameter of about five centimetres, with a circular or rectangular cross-section, and is about five metres long. It is usually constructed from brass or stainless steel and the diaphragms are plastic or metal foils. Shock velocities are obtained by measuring the time which the shock takes to pass a series of detectors. These are usually thin-film platinum resistance gauges, which respond to the change in temperature, or piezo-electric gauges, which respond to the change in pressure. An indication of the potential of the device may be gained by considering the results produced by a typical set of starting conditions. A driver pressure of 3 atm (300 kN m^{-2}) of helium combined with an experimental gas pressure of 1 Torr (133 N m^{-2}) of argon generates a shock wave moving at a velocity of 2200 m s^{-1} or Mach 6.6 i.e. 6.6 times the speed of sound, at a pressure of 55 Torr (7.3 kN m^{-2}) and a temperature of 4250 K. Because the temperature depends on the initial pressure ratio these high temperatures can be achieved even with glass equipment. Shock tubes have been used to generate temperatures as high as 20 000 K although, in the first kinetic study, the tube was cooled to $-35\,^{\circ}C$ and the shock used solely to return the gas to room temperature.

The shock tube provides a high temperature reactor of quite unique properties, but in order to employ it for kinetic studies a suitable detector must be chosen to monitor changes in the composition of the system. Quite severe constraints are imposed on the technique employed; obviously it cannot be allowed to protrude into the moving gas or to interfere in any way with the flow geometry. Because of the high speed of the shock wave and the consequent need to make all measurements within a short period of time the detector must be small in effective size. This usually implies a spatial resolution of about a millimetre and a time response of about a microsecond. It must be very sensitive, since the absolute quantities of material used are small, and for the most valuable results it should also be highly selective. The technique which meets these stringent criteria most successfully is optical spectroscopy. Absorption of visible and ultraviolet radiation provides a sensitive and selective method for measuring concentrations

and the photomultiplier-oscilloscope combination meets the required time and space resolution. A common experimental arrangement for the purpose is illustrated in Fig. 5.3.

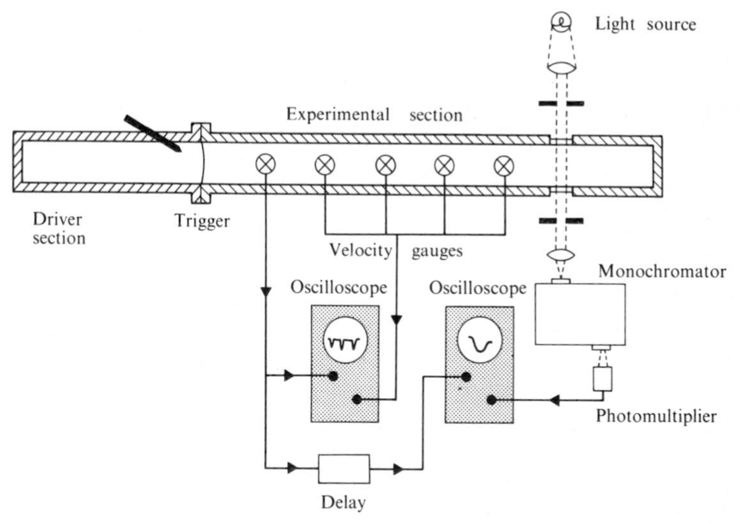

Fig. 5.3. Typical experimental layout with conventional shock tube.

When the shock wave reaches the end-wall of the tube it reflects as a shock, the moving gas behind being brought to a standstill and at the same time further heated. For the typical conditions quoted above, the reflected shock pressure would be 307 Torr (41 kN m^{-2}) and the temperature 9740 K. Although the details of this reflection process are by no means fully understood, the presence of this stationary gas extends the range of detection techniques which can be applied. Thus, mass spectrometry meets the requirements of sensitivity, selectivity, and time response but the need physically to transport reactant gas into the instrument virtually rules out the technique for use with the conventional shock tube using incident waves. However it is relatively straightforward to bleed stationary gas behind the reflected wave into the mass spectrometer (Fig. 5.4).

The third configuration is based on the *rarefaction* wave, formed simultaneously with the shock, which passes into the high-pressure section. Although this is not a step transition, in the same way as the shock, it behaves like a shock wave in reverse in that it leads to expansion and therefore cooling of the gas. The rate of cooling is typically

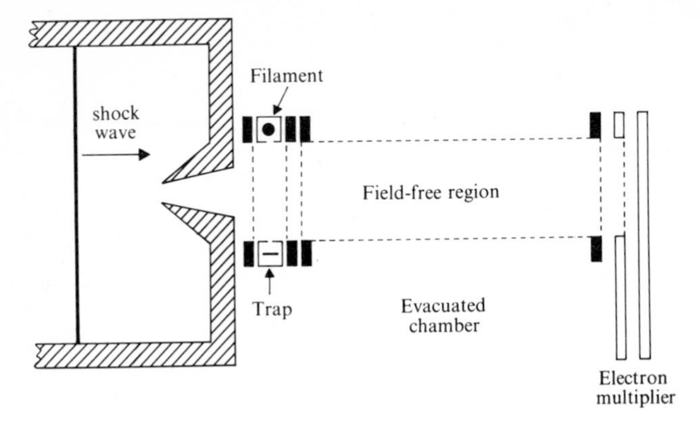

Fig. 5.4. Mass spectrometer sampling from the shock tube.

10^6 K s^{-1} and this reduction in temperature, coupled with a corresponding reduction in density, is adequate to quench many chemical reactions. By arranging two systems 'back-to-back' as shown in Fig. 5.5 it is possible to generate a cooling wave which will overtake the shock after an appropriate time interval. The configuration produces a heating pulse of known duration and enables the processed gas to be analyzed by more conventional means.

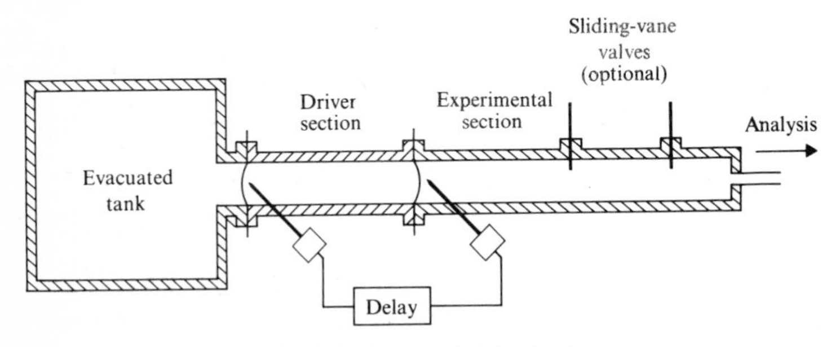

Fig. 5.5. Single-pulse shock tube.

There are few basic limitations of the shock tube technique. For practical purposes, the temperature range is unlimited and there is little serious restriction on the pressures employed although, below about 1 Torr, the shock front may become distorted. The upper limit on pressure is determined principally by the ability of the driver section to withstand high pressures. The greatest limitation is on the rate measurement itself: effective detector response below a microsecond becomes very difficult to achieve and above about a millisecond the advantages of the shock tube in terms of uniformity and homogeneity begin to be lost. The temperature and pressure conditions are therefore selected to provide measurable extents of reaction within these time limits.

The earlier measurements in shock tubes gave data with considerable scatter but the technique has now been sufficiently refined that rate constants of quite surprising accuracy are obtained with conventional shock tubes. Workers in independent laboratories using shock tubes of varying geometries equipped with different detection techniques can now obtain rate constants which agree to better than ± 50 per cent. For the reactions in question this is really quite remarkable. The quality of the data depends on the accuracy with which the velocities are measured, the signal/noise ratio of the detection technique, the method of analyzing the data and so on.

One problem peculiar to shock tubes must be mentioned. The gas behind the shock front is in motion with respect to the walls of the tube and, in consequence, a cold, stationary layer of gas builds up behind the moving front. Although the species in the layer do not react at the same rate as those in the *free stream*, this has a fairly minimal effect provided one is not monitoring residual reactant concentrations at high conversions. The greater problem is that the growth of the boundary layer causes the core of the gas stream to accelerate and with it the *contact surface.* The driver gas behaves rather like a 'leaky' piston. This affects the whole time-temperature history of the experimental gas. Although theoretical analysis of boundary layer growth is still being refined, it is sufficiently well developed to permit satisfactory corrections to be applied to data obtained from conventional tubes. However the whole aerodynamic process becomes exceedingly complex when the shock wave reflects since it returns through a gas stream with a core of changing temperature and velocity surrounded by a stagnant boundary layer. This means that the mass spectrometer sampling technique and the single-pulse shock tube tend to produce data of rather lower precision. Such techniques are by no means valueless but are more suited to studies of reaction mechanism and of relative, rather than absolute, rates.

Applications of the shock tube

The shock tube has been employed for the investigation of an enormous range of fast reactions but space limitations only allow mention of four areas of current interest.

The role of vibrational excitation in chemical reactions

We have already seen from the studies with crossed molecular beams

(Chapter 2) and with infrared chemiluminescence (Chapter 3) that an important question for the kineticist is the role played by the different energy modes in promoting chemical reaction. Most standard techniques for measuring reaction rates are of little value because energy exchange between the modes is much more rapid than the chemical reaction itself and a Boltzmann population of levels is maintained thoughout. This statement is less true of electronically-excited states and Chapters 4 and 6 show how the techniques of flash photolysis and of fluorescence quenching provide information on the reactions of triplet and singlet excited states respectively. In the majority of cases transfer between rotation and translation is so rapid that the question borders on the trivial; this still leaves open very important questions over the participation of vibrational motions.

The efficiency of vibrational energy transfer normally varies with temperature according to the relation $\exp(-1/T^{\frac{1}{3}})$, first predicted by Landau and Teller, whilst chemical reaction displays the common Arrhenius $\exp(E/RT)$ dependence. Thus by increasing the temperature it should always be possible to reach a situation where the two rates become comparable. If this also coincides with the typical time scale of shock tube measurements, say 5 - 500 μs, coupling between the processes should be observed. Such effects are most likely to be important for the stable diatomic molecules H_2, D_2, O_2, N_2, and CO which have relatively long vibrational relaxation times.

Perhaps not surprisingly the role of vibrational energy is found to be different in different reactions. The dissociation of oxygen above 11 000 K shows a distinct 'incubation' time, of the order of the vibrational relaxation time, during which no reaction occurs. Following this period, dissociation occurs at the rate predicted by a conventional Arrhenius relation for a quasi-equilibrium system. Nitrogen, on the other hand, displays no such effect either in its dissociation or in its reaction with oxygen atoms

$$N_2 + O \rightarrow NO + N$$

and in both cases the rate depends strictly on the translational temperature. By combining different techniques, it has now been convincingly demonstrated that the hydrogen-deuterium exchange reactions

$$D_2 + H_2 \rightarrow HD + HD$$

$$D_2 + NH_3 \rightarrow NH_2D + HD$$

$$D_2 + CH_4 \rightarrow CH_3D + HD$$

$$D_2 + H_2S \rightarrow HDS + HD$$

occur via specific vibrationally excited levels. Similar results have been obtained for isotopic exchange in oxygen and in nitrogen.

The rates of elementary reactions

A second major use of shock tubes lies in measuring the rates of elementary reactions. This is possible largely because at these high temperatures the stationary concentrations of atoms and radicals become comparable to those of the stable species and one can follow them directly using their characteristic absorption spectra. In the ideal situation, a labile molecule is used as a source of the atom or radical and, provided appropriate conditions are chosen, it will be rapidly converted to the required intermediate without any significant loss of the second reactant. This has been accomplished for O atoms using ozone as the source, for OH radicals using H_2O_2, and CH_3 using $CH_3N_2CH_3$. An alternative approach is to use a complex reaction sequence for which the mechanism is well established and select conditions under which an appropriate feature of the system becomes sensitive to the process of interest. This method has been used to determine the rates of the reactions

$$H + O_2 \rightarrow OH + O$$

$$O + H_2 \rightarrow OH + H$$

$$OH + CO \rightarrow H + CO_2$$

using the H_2-O_2 system and reactions of methane using the CO/O_2 system. The data analysis can be carried out satisfactorily only if the system is 'modelled' by computer because of the failure of the stationary state hypothesis and the complexity of the mechanism at high temperature.

The study of pyrolysis reactions

The shock tube is particularly suited to the study of pyrolysis reactions because the heating occurs entirely in the gas phase. The single-pulse shock tube enables any reactant to be investigated although the role of subsequent reactions is difficult to assess. Where spectroscopic or mass-spectrometric techniques can be employed to monitor the concentrations of intermediates and products, such an approach is to be preferred, and most studies on the simpler inorganic molecules have employed the conventional shock tube. In the case of hydrocarbons, subsequent reactions are more difficult to follow because available techniques fail to distinguish between the various products and the single-pulse shock tube is then equally as useful.

An enormous range of pyrolysis reactions has been studied and, in addition to a valuable collection of data, some interesting general features have emerged, not all of which can yet be fully explained. In the dissociation of a diatomic molecule the experimental activation energy is expected to equal the bond dissociation energy, although the contribution of additional degrees of freedom, other than translational motions along the line-of-centres, may affect the temperature dependence of the pre-exponential factor and hence an activation energy measured on the basis of a simple Arrhenius relation. In virtually every instance the activation energy turns out to be significantly lower than the endothermicity (by up to 90 kJ mol^{-1}) even though the overall rates are in agreement with theory. No satisfactory explanation has yet been given although it seems to be related to the coupling between vibrational relaxation and chemical reaction mentioned above, i.e. to the rate at which vibrational levels below the dissociation limit are populated.

In the case of triatomic molecules, the notion of simple bond rupture appears to be invalid, and instead prior formation of an electronically-excited intermediate is indicated. Thus sulphur dioxide is believed to decompose via an excited triplet state while CS_2 involves an excited singlet. Such decompositions are further complicated by subsequent chain processes. The problem becomes even more acute as the complexity of the molecule increases. Where it can be resolved, many reactions appear to obey the predictions of unimolecular reaction rate (RRKM) theory† and display the expected fall-off characteristics although there is evidence that the behaviour of some hydrocarbons may be considerably at variance with theory.

The study of combustion reactions

Self-supporting oxidation, or combustion, systems suffer from the presence of steep concentration and temperature gradients, which make the measurement of the elementary reaction rates involved very difficult. With the shock tube, virtually identical conditions of concentration and temperature can be achieved aerodynamically, the only difference being that a large excess of inert diluent must be added to minimize changes in the temperature and assist in its measurement.

Normally a spectroscopic technique is employed to monitor the changes in atom or radical concentration during the course of reaction. An assumed mechanism, in which the numerical values of the appropriate rate constants are varied, is then fitted to a series of such profiles obtained under a variety of conditions.

Certain features of combustion make it awkward to study, mainly on account of the high temperatures involved. In particular, rates of reaction are less dependent on their associated activation energies and, because of

† See Pilling's *Reaction Kinetics* (OCS 22) for a brief introduction.

the high radical concentrations present, radical-radical processes become important. The net effect is that, for a given system, more possible reactions become significant than at lower temperatures. Added to this, the branching-chain nature of the system means that most of the experimental observations turn out to be closely interrelated and many of the species concentrations remain in steady-state with each other.

In the traditional hydrogen-oxygen reaction, observation of the OH profile reveals the following phases of reaction (Fig. 5.6):

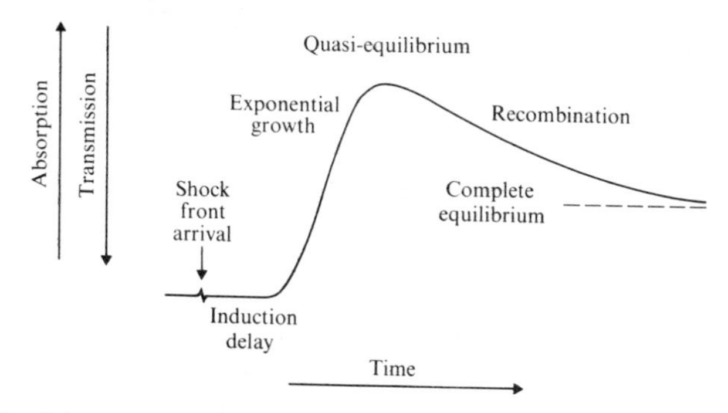

Fig. 5.6. Concentration-time profile for the hydroxyl radical during the hydrogen-oxygen reaction.

A. Initiation and branching phase

$$H_2, O_2 \rightarrow H, O, OH$$
$$H + O_2 \rightarrow OH + O$$
$$O + H_2 \rightarrow OH + H$$
$$OH + H_2 \rightarrow H_2O + H$$

B. Quasi-equilibrium phase (shuttle reactions)

$$H + O_2 \rightleftharpoons OH + O$$
$$O + H_2 \rightleftharpoons OH + H$$
$$OH + H_2 \rightleftharpoons H_2O + H$$

C. Termination (or exothermic) phase

$$H + H + M \rightarrow H_2 + M$$
$$H + O + M \rightarrow OH + M$$
$$OH + H + M \rightarrow H_2O + M$$

etc.

By careful choice of experimental conditions it is possible to isolate the separate reactions and determine extremely accurate rate constants.

With more complex systems e.g. $CO - O_2$, $CH_4 - O_2$ and $C_2H_2 - O_2$, the picture becomes much more complex although the shock tube probably remains the most suitable technique available for their study.

6. Low intensity photolysis

Fluorescence quenching

IN Chapter 4 it was shown that a molecule which absorbs light is raised initially to an electronically-excited singlet state. The excited molecule then falls rapidly to the lowest vibrational level and, if it does not undergo any other process in the meantime, it *fluoresces* or re-emits radiation, at a wavelength somewhat longer than that it absorbed. Because the lifetime of a singlet state is short, typically 10^{-8} s, the majority of molecules can be expected to fluoresce in this way. However, if other processes are able to remove energy within this period, the excited molecules will be *quenched* and the radiation intensity reduced. The basis of the technique is to study the fluorescence intensity as a function of concentration of the quenching agent. It may thus be categorized as a competitive method in which a physical process, in this case emission of radiation, competes with the chemical process to produce a steady-state situation.

In the simplest case the molecules are excited at a rate proportional to the intensity of the incident radiation

$$A + h\nu \xrightarrow{\phi} A^*$$

and are de-excited by one of three processes:

(1) Fluorescence to the ground state

$$A^* \xrightarrow{k_f} A + h\nu.$$

(2) Internal quenching due to interaction of the excited molecules with ground-state species or with solvent molecules

$$A^* \xrightarrow{k_i} A.$$

(3) External quenching due to reaction of the excited molecules with added reactant

$$A^* + R \xrightarrow{k_q} \text{products}.$$

In the steady-state

$$\frac{d[A^*]}{dt} = \phi[A] - k_f[A^*] - k_i[A^*] - k_q[A^*][R] = 0. \qquad (6.1)$$

The fluorescence intensity I, equal to $k_f[A^*]$, is then given by

$I = \dfrac{k_f \phi [A]}{k_f + k_i + k_q [R]}$. In the absence of added reactant, the corresponding intensity I_o is given by the same relation but with the omission of the final term in the denominator. The ratio of the two intensities is thus

$$\frac{I_o}{I} = \frac{k_f + k_i + k_q [R]}{k_f + k_i} = 1 + \frac{k_q [R]}{k_f + k_i}. \tag{6.2}$$

This is known as the *Stern-Volmer equation* and a plot of $1/I$ versus [R], which should be linear, is known as a *Stern-Volmer plot.* The technique is readily used to obtain relative quenching efficiencies of various reactants simply by comparing the slopes of the appropriate Stern-Volmer plots.

The determination of absolute rates is more difficult since the sum $k_f + k_i$ has to be found although the measurement needs to be made only once for any molecule-solvent combination. Several methods are available and two are worth mentioning. The most obvious is to irradiate the sample with a light pulse of very short duration, normally less than 10^{-9} s, and then follow the decay of the fluorescence after the extinction of the pulse by means of a photomultiplier-oscilloscope combination. This is reminiscent of the nanosecond flash photolysis method (p. 40), although, since the emission is relatively intense, the signal/noise problem is reduced. Another approach is to use a modulated light source and examine the phase shift of the resulting emission. This method possesses the advantages characteristic of steady-state measurements.

The derivation of the Stern-Volmer equation given above is perfectly adequate for gaseous systems, but in solution the situation is frequently more complex. The reason for this is that the quenching reactions are necessarily fast, and in consequence likely to be diffusion-controlled. In such a situation two effective quenching mechanisms may be distinguished, one which arises when the quencher has to diffuse into the solvent 'cage' (the solvent in the immediate vicinity of the excited molecule, not necessarily a rigid structure), and the other when the quenching molecule is already present in the 'cage' and is therefore associated with the molecule at the moment of excitation. It is not worthwhile here to describe the more complex treatment required to deal with such situations; however, one can predict qualitatively that at low quencher concentrations the normal Stern-Volmer behaviour will be observed, with the quenching rate dictated by diffusion. At higher concentrations more of the excited molecules are associated with the quencher and so are immediately de-excited. The Stern-Volmer plot is then no longer linear, but displays curvature towards high quencher concentrations.

There is no need to dwell on the experimental arrangements since the apparatus is quite straightforward and commercial instruments are available. Basically it is a matter of irradiating the sample cell with a suitable monochromatic source and observing the fluorescence by means of photomultiplier placed at right angles to the incident beam. The experimental difficulties are usually associated with the preparation of the sample, owing to the necessity of excluding trace impurities which may themselves interact with the light or act as quenchers.

The scope of the fluorescence technique is more limited than most others described in this book since it applies only to the excited states of a limited number of molecules, and the effective rates must lie within a fairly narrow range (10^8 to $10^{12}\,s^{-1}$). On the other hand it is simple to use and is virtually the sole technique available for these particular reactions.

Because relative rates are readily obtained and the effects of solvent, viscosity, and temperature can be established quite rapidly, the technique has proved very useful for studying diffusion control of the quenching reactions. It has also been used to study quenching mechanisms, some of which prove to be very interesting. *Self quenching* by dimer formation is one phenomenon observed; in the case of anthracene, collision between the excited state and the ground state immediately quenches the fluorescence but, with pyrene, the dimer itself fluoresces (though at rather longer wavelengths). Another observation is that energy transfer can occur over surprisingly large distances, the critical separation for quenching of acriflavine by rhodamine B being 7 nm (70 Å). It is also worth noting that proton transfer reactions, which have been subjected to study by many other methods described in this book, can be monitored by this technique if the reactant is an aromatic species for which the acidic and basic forms fluoresce at different wavelengths.

Modulation techniques

The rotating-sector method

The rotating-sector technique constitutes one of the simplest and most elegant methods of studying fast reactions. Its principal use has been in the investigation of photo-induced addition polymerization although it is certainly not restricted to such reactions.

A simple photo-initiated chain reaction, in which a reactant A is transformed into a product B, may be represented schematically by three processes:

Initiation, I	Reactant, A, or Initiator $\xrightarrow{h\nu}$ 2R·
Propagation, k_p	R· + A $\rightarrow$ B + R·

Termination, k_t $\quad R\cdot + R\cdot \rightarrow R_2$

When a steady-state has been established, the stationary radical concentration is given by

$$\frac{d[R\cdot]}{dt} = 0 = I - k_t[R\cdot]^2 \tag{6.3}$$

and the overall reaction rate by

$$-\frac{d[A]}{dt} = \frac{d[B]}{dt} = k_p[R\cdot][A] = \frac{k_p I^{\frac{1}{2}}[A]}{k_t^{\frac{1}{2}}}. \tag{6.4}$$

Conventional kinetic measurements yield the composite quantity $k_p/k_t^{\frac{1}{2}}$ and the individual components k_p and k_t cannot be determined simply by altering the concentration, light intensity, etc. One way of overcoming the problem would be to measure the stationary radical concentration, but as we have already seen (Chapter 4), this can be accomplished only under rather extreme conditions. However, separation of the two quantities becomes possible if establishment of the steady state can be prevented.

A mean lifetime τ for the radicals may be defined under steady-state conditions as the concentration of radicals divided by their rate of removal. For the situation above

$$\tau = \frac{[R\cdot]}{k_t[R\cdot]^2} = \frac{1}{k_t[R\cdot]}. \tag{6.5}$$

Combining with (6.4) gives

$$-\frac{d[A]}{dt} = \frac{k_p}{k_t} \cdot \frac{[A]}{\tau}.$$

Thus a measurement of the lifetime τ, taken together with the overall steady-state propagation rate, yields a measure of the ratio k_p/k_t and hence permits evaluation of the two separate rate constants.

Integration of the kinetic equations shows that the radical concentration relaxes to its steady state value with a relaxation time closely related to the mean lifetime τ. If, instead of irradiating continuously, the light is switched on for a fraction F of the total duration of the experiment t_{tot} and each period of illumination considerably exceeds τ, the system behaves as if it were receiving the full intensity I of the light but only for a reaction time Ft_{tot}. This is represented by curve (a) in Fig. 6.1.

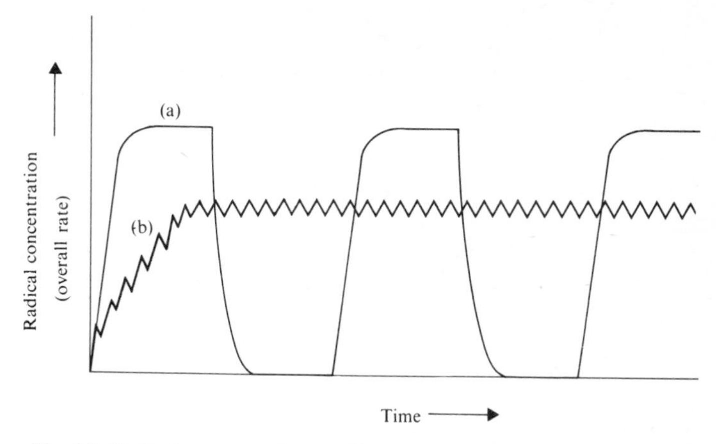

Fig. 6.1. Radical concentration profiles using the rotating-sector method.

However, if the light periods are considerably shorter than the lifetime τ, the radical concentration follows curve (b) (Fig. 6.1). In this case the system behaves as if the illumination lasted for the complete duration of the experiment t_{tot} but at a reduced intensity $F\,I$. From eqn (6.4) the overall rate of reaction depends on the square root of the intensity so that as the duration of the light periods is reduced the extent of reaction changes from $F\,R_o$ to $F^{\frac{1}{2}}\,R_o$, where R_o is the extent of reaction for continuous illumination.

A plot of the result obtained is shown in Fig. 6.2. The turning-point t' on the curve is closely related, although not exactly equal, to the mean lifetime τ. The theory of the technique has been developed to a considerable degree of sophistication, with allowances for non-square light pulses, non-uniform illumination, interference by 'dark' reactions, and complex termination sequences. The computations have been tabulated so that the measured values of t' can be transformed to corresponding values of τ and hence to the required rate constants.

The apparatus needed is also quite simple. Light from a suitable source, usually a mercury arc, is brought to a focus and the rotating-sector inserted at this point. The slots occupy about one-quarter of the total so that the light periods have one-third the duration of the dark periods. The sector is driven by a synchronous motor to give close control over the speed of rotation. The light is made parallel by a simple lens system and then is allowed to fall on the reaction cell.

The technique is applicable whenever a reaction rate is not first-order in light intensity. Termination rate constants with values in the range

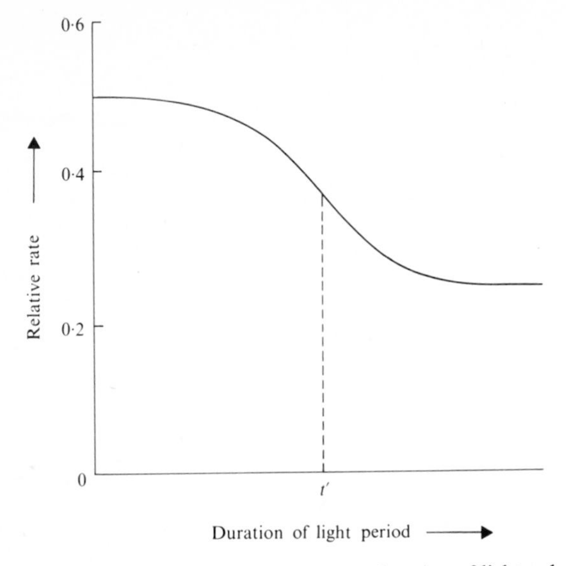

Fig. 6.2. Dependence of reaction rate on duration of light pulse.

10^7 to 10^{10} $l\ mol^{-1}\ s^{-1}$ have been measured in this way. These are fast, bimolecular reactions and the onset of diffusion control is readily observed as the viscosity of the solvent is raised.

Another interesting application has been in the bimolecular recombination of iodine atoms in solution, where the competing first-order reaction was created by introducing a radioactive tracer and following the rate of the exchange reaction

$$I + I^*I \rightarrow I^* + I_2.$$

The gas phase recombination of methyl radicals, which is important in all gaseous organic systems, has also been investigated in this way using the photolysis of acetone as the radical source. Above 400 K with 250 nm radiation, the reaction scheme is simply

$$CH_3COCH_3 \rightarrow 2CH_3 + CO$$

$$CH_3 + CH_3COCH_3 \rightarrow CH_4 + CH_2COCH_3$$

$$CH_3 + CH_3 \rightarrow 2C_2H_6$$

$$CH_3COCH_2 + CH_3COCH_2 \rightarrow (CH_3COCH_2)_2$$

$$CH_3 + CH_3COCH_2 \rightarrow CH_3COCH_2CH_3.$$

The production of methane was used to monitor the propagation rate, the lifetime of the methyl radical being determined by the first of the three termination reactions. The result $k = 5 \times 10^{10} \exp(0 \pm 3 \text{ kJ mol}^{-1}/RT)$ l mol^{-1} s^{-1}, has been confirmed by using the photolysis of dimethyl mercury as an alternative source.

Phase shift measurements

In the rotating-sector technique the reaction rate is studied as a function of the chopping frequency, the basis of this approach being that the rate is proportional to the amplitude of the fluctuating radical concentration. A more detailed analysis shows that the phase of the radical concentration also varies with the frequency f of the exciting radiation according to the relation

$$\tan \delta = 2\pi f \tau .$$

The principal advantage of this approach is that phase-sensitive detection and amplification can be employed and the phase-shift and absorption spectra of transient species obtained, even when the signal strength is below the random noise level. This means that spectra of transient species of the kind normally associated with flash photolysis can be obtained with low intensity photolysis.

The measurement of phase shifts as a way of obtaining fluorescent lifetimes has been referred to earlier in the chapter. Further examples of the use of the method are the separation of the ultraviolet spectra of ClO and ClO_2 by means of their different phase shifts and the detection of the infrared absorption of ClO_2. A particularly important example is the identification of both the infrared and ultraviolet spectra of the hydroperoxy radical HO_2. A great deal of the theory of gas-phase oxidation depends on the properties of this radical, yet it is only recently that it has actually been observed directly.

7. Flow techniques

THE earliest methods developed for the direct measurement of fast reaction rates employed flow techniques. Hartridge and Roughton introduced these techniques in 1923 as means of studying biochemical processes in liquids, and the same approach was used by Paneth in 1929 to establish the existence and reactivity of gaseous free radicals. Despite the subsequent introduction of more sophisticated methods, flow techniques still remain in extensive use fifty years later.

In its simplest form, the *continuous flow* method provides an excellent example of the way in which competition between a chemical reaction and a physical process (in this case steady flow along a uniform tube), avoids the necessity of working in real-time and allows measurements to be made at leisure. The principle has already been described in Chapter 1 (see Fig. 1.3). Two liquids, which react spontaneously, are allowed to mix at one end of the flow tube and then travel along it at uniform velocity. This velocity can be determined simply by measuring the volume of fluid collected at the exit over an appreciable time interval. Since reaction is occurring during the flow, the concentrations at different positions along the tube correspond to different reaction times, and the time co-ordinate is thus replaced by a distance co-ordinate.

A serious drawback of the continuous flow method is the large volume of fluid consumed. This is rarely a problem in gaseous systems but becomes very serious with biochemical reactions, where the quantity of pure, biologically-active material available is likely to be very small. For this reason the subsequent development of these techniques for use with liquid systems has differed from that for gases.

Flow methods for liquid systems

Continuous-flow methods

The ultimate limitation on time resolution in all flow systems arises from the finite time required for mixing. This time can be kept to a minimum by ensuring that the flow is highly turbulent in the mixing chamber, usually by introducing the liquids tangentially at high velocity through small jets. Mixing times achieved in this way are about a millisecond, which is equivalent to a distance of 1 cm for a flow velocity of $10\ \mathrm{m\,s^{-1}}$.

The rate measurements are normally made on the assumption that *plug flow* is realized i.e. that all fluid elements move with the same velocity. In practice fluid adjacent to the wall moves at a lower velocity than that on the axis so this approximation must be examined further.

In the most adverse case, the flow becomes *streamline* or *laminar* and the velocity profile is parabolic with the fluid on the axis moving at double the mean velocity. If the flow is highly *turbulent*, fluid elements on the centre-line exchange rapidly with those on the periphery and the radial velocity distribution becomes flattened. A criterion for transition between laminar and turbulent flow is provided by the dimensionless *Reynold's number* (Re), given by $(u \times l)/\bar{\nu}$ where u is the velocity, l is a characteristic dimension, normally the diameter, and $\bar{\nu}$ is the kinematic viscosity, i.e. the viscosity divided by the density. The transition usually occurs in the range $(Re) \sim 2300 - 3200$ so that a Reynold's number above this value is to be preferred. In practice, water flowing through a 1 cm diameter tube becomes turbulent at velocities above about 30 cm s^{-1} so that the condition is quite easily achieved. The error introduced by treating turbulent flow as plug flow does not normally exceed 5 per cent.

The volume limitation referred to above is not trivial; early experiments required several litres of reagent per experiment which placed considerable restriction on systems of biological interest. Recent apparatus has employed small-bore tubing, down to 1 mm diameter, generating the flow by means of mechanically-operated hypodermic syringes, and rapid-scanning spectrophotometric detection. With such methods, the volume required can be reduced to a few millilitres.

Spectrophotometric detection has been most commonly used with this technique, as with many others described in this volume, but conductivity measurement has also proved very useful. Of greater interest in terms of novelty is the monitoring of reaction progress by heat release. This employs a movable thermocouple in the centre of the flow tube. With careful thermostatting, temperature changes of 0.03 °C can be measured with an accuracy of a few per cent. Using decimolar solutions, reactions with heat evolutions as low as 1 kJ mol^{-1} can be investigated.

Accelerated-flow and stopped-flow methods

The distinctive feature of the continuous flow technique is that it is a stationary-state method and the time response of the detector is unimportant, so that inherently slow detection techniques can be employed, the penalty being the fluid volume required. If the detection technique has an adequate time response, it is therefore preferable to conserve the reagents by making real-time measurements. The 'flow' aspect of the method is still required to ensure mixing in times less than those required for reaction.

In the first alternative–the *accelerated flow* method–the observation station is situated close to the mixing chamber and recording is commenced as the pistons start to move. The flow rate increases steadily as the plungers accelerate so that, at the detection station, the overall reaction time decreases. This has the rather curious result that the record obtained is a reverse of the usual progress curve i.e. the reactant concentration appears to increase with time. The observation does not correspond directly to real-time because the flow speed increases during the experiment and it is necessary simultaneously to monitor the movement of the pistons to effect the conversion.

This technique has largely been superceded by the *stopped-flow* method in which steady flow is first established and is then brought rapidly to a halt, after which the subsequent concentration changes are monitored in real-time. Fast flow through the mixing chamber is needed to ensure adequate mixing before significant reaction has occurred; this flow must be halted very rapidly in order to avoid simply observing incompletely-mixed reactants. This cessation of flow is normally achieved by the simple expedient of making the liquid drive a movable piston which is brought up against an external stop (see Fig. 7.1). Typically the

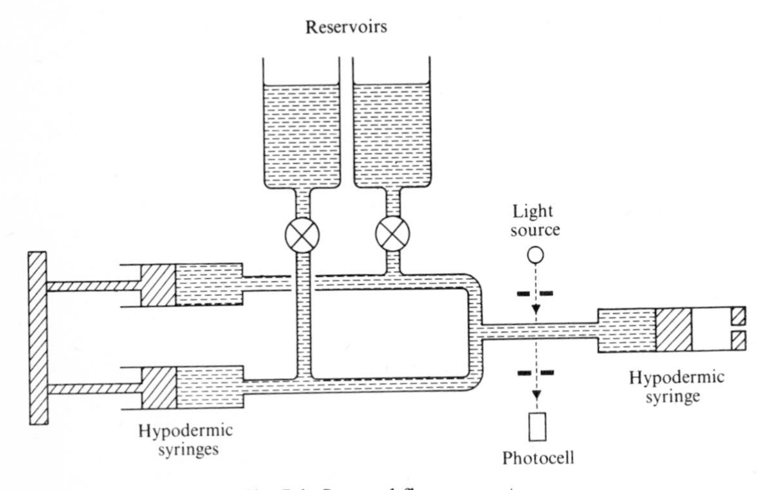

Fig. 7.1. Stopped-flow apparatus.

hypodermic syringes would have a capacity of *ca.* 2 ml, the observation tube a diameter of *ca.* 2 mm with measurements made *ca.* 8 mm from the mixing chamber. Such a device permits kinetic studies to be conducted on as little as 0.15 ml of each reactant with reaction times down to 2 or 3 milliseconds.

This technique has now become quite standardized and is in fact one of the few fast reaction techniques which has been embodied in a commercial instrument.

Application in enzyme kinetics

The outstanding application of flow techniques has been in the study of the kinetics of enzyme reactions. The theoretical basis of enzyme kinetics was first put forward by Michaelis and Menten in 1913. They proposed that enzyme action could be explained by a simple two-step process

$$\mathrm{E + S \underset{k_2}{\overset{k_1}{\rightleftharpoons}} ES \overset{k_3}{\rightarrow} P + E}$$

where E is the enzyme, S is the substrate, ES the enzyme-substrate complex and P the product. Under steady-state conditions the rate of formation of P is given by

$$\frac{\mathrm{d[P]}}{\mathrm{d}t} = \frac{k_3[\mathrm{E}]_0}{1 + K/[\mathrm{S}]}$$

where $[\mathrm{E}]_0$ is the total (or initial) enzyme concentration, $[\mathrm{E}] + [\mathrm{ES}]$, and K is the Michaelis-Menten constant, $(k_2 + k_3)/k_1$. (Readers should recognize the similarity between this relation and that proposed by Lindemann to explain the behaviour of first-order reactions.)† Conventional kinetic studies yield the value of k_3 and of the Michaelis-Menten constant but do not permit evaluation of the individual rate constants k_1 and k_2.

The improved time response of flow techniques has made it possible to monitor the growth and decay of the enzyme-substrate complex ES and hence the kinetics of the elementary processes involved. Perhaps more important, such investigations provided a direct demonstration of the Michaelis-Menten theory and brought out examples where the theory, although basically correct, required further refinement to explain the results.

An example is provided by the studies of the kinetics of action of catalases and peroxidases: the *catalatic* action involves the reaction with hydrogen peroxide to give oxygen and water and the *peroxidatic* reaction causes oxidation of a donor molecule in the presence of a peroxide. The kinetics involve a second-order reaction of the enzyme-substrate complex and the Michaelis-Menten formalism must be modified accordingly

$$\mathrm{E + S \rightleftharpoons ES}$$

$$\mathrm{ES + AH_2 \rightleftharpoons E + SH_2 + A.}$$

† See Pilling, *loc. cit.*

In this case, the peroxide acts as the substrate S and AH_2 represents the donor molecule. In the catalatic reaction involving hydrogen peroxide, the donor and substrate are identical so that the mechanism simplifies to

$$E + S \rightleftharpoons ES$$

$$ES + S \rightleftharpoons E + P.$$

In both cases, the Michaelis-Menten 'constant' is no longer a true constant but involves the donor concentration $[AH_2]$. Also, it is found that these enzymes form a sequence of different enzyme-substrate complexes identifiable by their colour. In the case of the peroxidases both the first two complexes formed react with donor molecules although at markedly different rates.

The capacity-flow, or stirred-flow, method

An alternative to the continuous flow configuration involves passing the reactants together into a large chamber which is very rapidly stirred (Fig. 7.2). Provided the time required for mixing is much shorter than

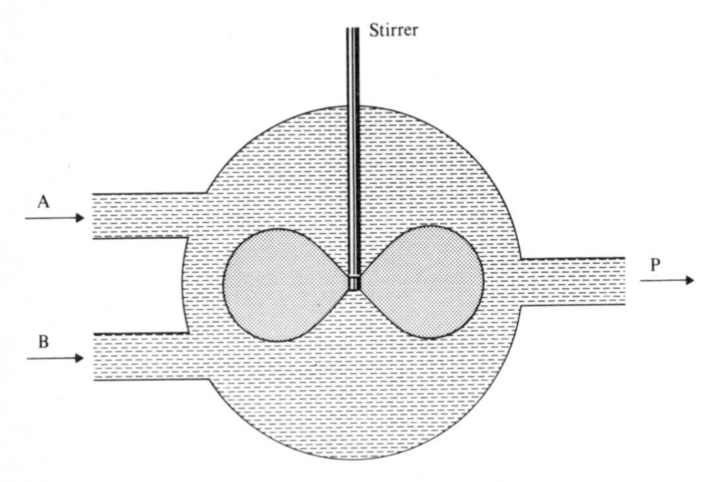

Fig. 7.2. Schematic representation of stirred-flow reactor.

the characteristic reaction time, the reactants will be distributed uniformly throughout the chamber. A steady-state situation will then be established in which the flow of reactants into the chamber is exactly balanced by the flow of products plus unchanged reactants out of it. Furthermore, if analysis of the exit flow is carried out close to the chamber, the measured concentrations will equal those in the chamber proper.

The evaluation of the rate data may be appreciated by considering the simplest possible reaction

$$A + B \rightarrow P.$$

The total rate in a vessel of volume V will be $V\,k\,[A]\,[B]$. The net rate of change of each species within the chamber, equal to zero in the stationary state, is obtained by combining the appropriate flows in and out with the rate of change due to reaction

$$V\frac{d[A]}{dt} = 0 = u_A[A]_o - Vk[A][B] - u[A]$$

$$V\frac{d[B]}{dt} = 0 = u_B[B]_o - Vk[A][B] - u[B]$$

$$V\frac{d[P]}{dt} = 0 = Vk[A][B] - u[P].$$

Summing these equations gives

$$[P] = \frac{V}{u^3}\,k\,\{u_A[A]_o - u[P]\}\{u_B[B]_o - u[P]\}.$$

Provided the volume change is negligible, $u_A + u_B = u$, so that apart from the flows entering the reactor it is necessary only to measure the product concentration in the efflux to obtain the rate constant k.

This simple, elegant technique is limited by the efficiency with which the reactants can be mixed. Very high stirrer rates, up to 15 000 r.p.m., prove necessary and residence times below 1 s are inaccessible. For gases at low pressures, however, diffusion suffices to ensure rapid mixing. One particular investigation employed a vessel of 200 ml capacity at a pressure of 270 N m^{-2} (2 Torr) and under these conditions, the time for an active species to traverse the vessel was 0.04 s. The stirred reactor principle is applicable to any residence time in excess of this figure. In the study referred to, the residence time was 0.35 s and the mixing time was reduced even further by admitting one of the reactants at high linear velocity.

The stirred reactor principle has been utilized only in a limited number of cases, for example, the hydrolysis of esters and the bromination of acetone, although it appears ideal for systems which fall just below the time scale involved in conventional measurements.

Flow methods for gaseous systems

Discharge-flow methods; measurement of elementary reaction rates

Very few stable gases react together spontaneously at an appreciable rate and the value of gas-phase flow systems lies in the opportunity of producing atoms and radicals by means of radiofrequency or microwave discharges at low pressures. Some of the species which can be generated in this way are listed in Table 7.1.

TABLE 7.1

Gas-phase atoms and radicals detected by electron spin resonance†

H	Te	O_2	NS
D	F	NO	SF
^{14}N	Cl	^{15}NO	SeF
^{15}N	Br	OH	ClO
P	I	OD	BrO
As	Na	^{17}OH	IO
^{121}Sb	K	^{17}OD	NF
^{123}Sb	^{85}Rb	SH	CF
O	^{87}Rb	SD	NCO
S	Cs	SeH	NCS
Se	Ar	SeD	NO_2
		TeH	NF_2
		SO	HCO
		SeO	e_g^-

† Electronically-excited species have not been recorded separately.

Because of the nature of the species involved, the method adopted for following the progress of reaction now assumes a dominant role. Atom concentrations have been measured by the heat evolved during their recombination on a catalyst surface and by other physical measurements but, in general, a more selective technique is required. Those principally in use at present are electron spin resonance, molecular-beam mass spectrometry, and optical absorption.

Electron spin resonance. With this technique the flow tube passes through the microwave cavity and the time variation is achieved by movement of the complete magnet-cavity assembly, by movement of the mixing region, using an internal inlet probe, or by altering the flow velocity. Of these the movable probe usually proves the simplest and most reliable although in certain instances it can introduce variations in the nature of the surface. The detector is both sensitive and selective and does not interfere with the flow; it can also be used for studying reactions of free electrons using their cyclotron resonance absorption. The major limitation is that it can be used only for atoms and a limited range of diatomic radicals which give rise to electric dipole transitions (e.g. OH, SH, NO, SO). It is unlikely to be used successfully with polyatomic radicals because of the complexity of the spectra involved.

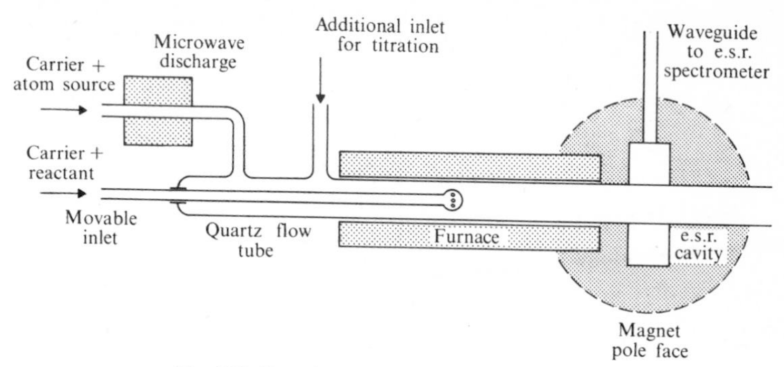

Fig. 7.3. Gas-phase discharge-flow apparatus.

Mass spectrometry. This technique is more difficult to employ because the gas has to be fed into the source through a special nozzle. Earlier studies suffered through heterogeneous reactions associated with the sampling process but *molecular-beam inlets* which convert a section of the mass flow into collisionless molecular flow largely circumvent the problem. The remaining difficulty, inherent in mass spectrometry, arises because overlapping spectra make data analysis more cumbersome. This problem can be reduced by working at lower ionizing energies obtained, for example, by using a photoionization source. In compensation, mass spectrometry possesses the very real advantage of detecting virtually all species present.

Optical absorption. Spectrophotometric detectors based upon optical absorption figure prominently throughout this volume. As a general rule, optical absorption is rather less satisfactory than the other two techniques; many species either do not absorb sufficiently or do so in a spectral region where other components interfere.

In gas-phase systems, the Reynold's number is usually very low and the flow is strictly laminar, with a parabolic velocity distribution, which does not even remotely approach the plug flow situation. Fortunately the low pressures used allow radial diffusion to remove the concentration gradients across the tube so that the longitudinal gradients correspond closely to those arising from plug flow. In consequence, however, back diffusion of reactants, at low velocities, and viscous pressure drop, at high flow velocities, introduce unavoidable corrections. The most serious problem in practice normally turns out to be the importance of wall reactions; atoms and radicals are rapidly destroyed at the surface of the tube and radial diffusion then fails to remove the transverse concentration gradient. Tubes are usually constructed from quartz and treated with

HF, boric acid, etc. to minimize the problems but appreciable effects often remain, in many cases, due to contamination by the reactants.

The major contribution of fast flow discharge systems has been in the measurement with high precision of rate constants of elementary reactions. Suitable examples are the studies using e.s.r. detection of OH radical reactions, which were the first to be monitored in this way. The elementary reactions involved were

$$OH + OH \rightarrow H_2O + O$$

$$OH + CO \rightarrow CO_2 + H$$

$$OH + H_2 \rightarrow H_2O + H$$

$$OH + CH_4 \rightarrow H_2O + CH_3.$$

These are all very fast, important in gas-phase combustion, and difficult to investigate in other ways, although an accurate knowledge of their rates is essential for a complete understanding of combustion processes. It is worth commenting that these reactions have very low activation energies, in contrast to earlier predictions.

Premixed flames

Flames are examples of gaseous systems in which flow motion competes with self-supporting chemical reaction to produce a steady-state situation. The proximity of the burner surface serves to stabilize the flame and, although common flames are highly distorted, special designs such as the Egerton-Powling burner illustrated in Fig. 7.4 generate a flat flame which can be studied by spectroscopic techniques and by sampling probes. Reaction rates have been obtained from flame studies but the steep concentration and temperature gradients make the data difficult to interpret. The concentration gradients turn out to be rather insensitive to the numerical values of many of the rate constants involved and to depend rather more critically on the physical characteristics of the system.

Despite the valuable results which have been extracted it is fair to say that a flame would not normally be selected for measuring a reaction rate unless the reaction in question had some particular association with self-supporting combustion. Thus, flames have been used to study chemiluminescent and chemi-ionization phenomena to great effect and it is largely as a result that excitation processes in hydrocarbon-oxygen systems are now becoming understood.

Chemiluminescence has been attributed to a large variety of reactions; the following are examples for three of the common emitters:

$$C_2 + OH \rightarrow CH^* + CO$$

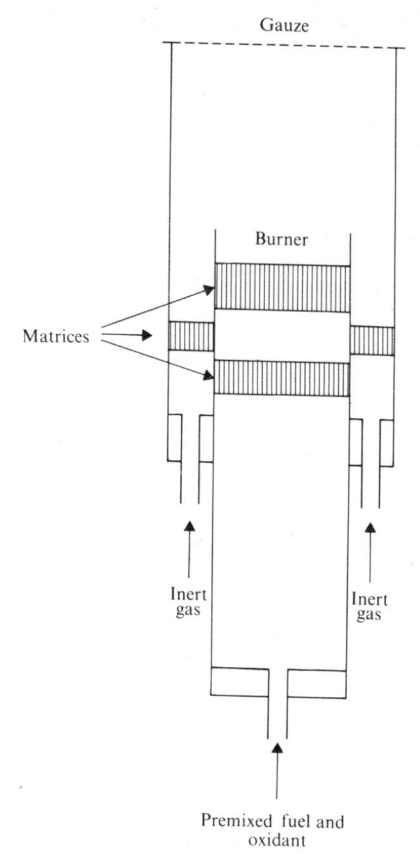

Fig. 7.4. Egerton-Powling burner.

$$C_3H_3^+ + e^- \rightarrow C_2^* + CH_3$$
$$CH + O_2 \rightarrow OH^* + CO.$$

It seems likely that most of the reactions postulated occur to some extent and it is difficult to establish their relative importance. Chemi-ionization, on the other hand, seems rather different, the major reaction now being clearly established as

$$CH(X, {}^2\Pi) + O \rightarrow CHO^+ + e^-.$$

The other ions observed, such as H_3O^+, which considerably exceed the stationary CHO^+ concentration, are formed by charge exchange between CHO^+ and neutral species.

Diffusion flames

An account of fast reaction techniques would be incomplete without a mention of diffusion flames used originally in the study of reactions of alkali metal atoms. Although the method is no longer employed, it was the first to provide rates of elementary reactions in the gas phase.

In contrast to the premixed flames described above, the two reactants are initially separate and diffuse into one another at the burner. The chemical reaction competes once again with the physical process of diffusion and the distance a reactant diffuses before it is consumed will be related (inversely) to the reaction rate.

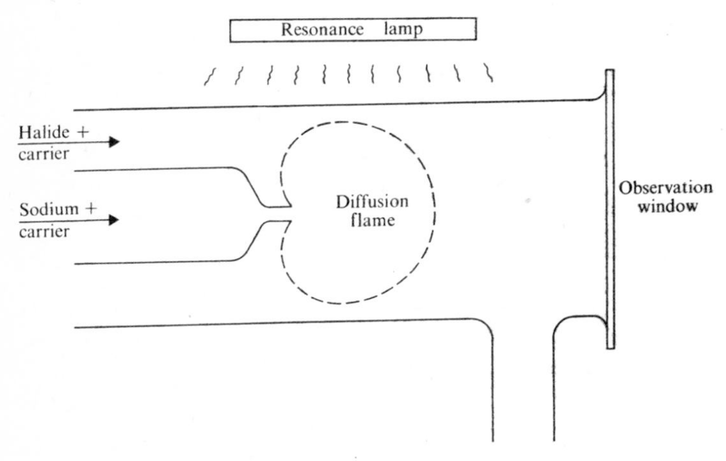

Fig. 7.5. Diffusion flame apparatus.

The experimental arrangement is illustrated in Fig. 7.5. A stream of carrier gas, at a pressure of 1 - 10 Torr, saturated with alkali metal vapour, flows through a nozzle, two or three millimetres in diameter, into a second stream containing a known concentration of an organic halide. A reaction zone, roughly spherical in shape, within which the process

$$M + RX \rightarrow M^+X^- + R$$

occurs, is situated at the nozzle tip. This region can be rendered visible to an observer by illuminating at right angles with the appropriate atomic resonance radiation. The limit of detectability, which defines the zone boundary, can be determined in independent experiments. A technique for locating the position of the boundary by measuring heat evolution with a small thermocouple has also been developed.

The major drawbacks of the technique, apart from obvious measurement problems, are that a complete description of the flow behaviour is mathematically intractable and that temperature variation is impractical because the temperature dependences of the detectability limit and of the diffusion coefficient are difficult to establish. Despite this, a large amount of valuable data on these reactions has been compiled. Based on the relative reactivities it was possible to demonstrate how negative substituents, such as additional halogen atoms, increased the reaction rate because of the polar nature of the reaction e.g.

$$Na + CH_2BrCl \rightarrow Na^+Cl^- + CH_2Br.$$

8. Direct relaxation methods

RELAXATION is a term which is used in a quite general sense to denote the approach of a system to its equilibrium position. It is also used to characterize a range of *relaxation techniques* which depend for their success on an important principle: this is, that when a system suffers only a very small displacement from equilibrium, the rate at which it returns is directly proportional to the magnitude of the displacement. Such a process is said to be *linear.* This may be expressed mathematically, representing the displacement as Δx, by

$$-\frac{\mathrm{d}\,\Delta x}{\mathrm{d}t} = \text{constant} \times \Delta x = c\,\Delta x$$

which integrates to

$$\Delta x = \Delta x_{\mathrm{o}}\,\mathrm{e}^{-ct} \text{ or } \Delta x_{\mathrm{o}}\,\mathrm{e}^{-t/\tau}$$

where Δx_{o} is the value of Δx at $t = 0$, and $\tau = 1/c$.

The quantity τ is the time required for the displacement to fall to 1/e of its original value; it is termed the *relaxation time* and is simply the reciprocal of the overall first-order rate constant for the process.

The simplest possible equilibrium may be represented by

$$\mathrm{A} \underset{k_2}{\overset{k_1}{\rightleftharpoons}} \mathrm{B}.$$

At equilibrium the forward and backward rates are balanced so that

$$-\frac{\mathrm{d}[\mathrm{A}]}{\mathrm{d}t} = \frac{\mathrm{d}[\mathrm{B}]}{\mathrm{d}t} = 0 = k_1\,[\mathrm{A}]_{\mathrm{o}} - k_2\,[\mathrm{B}]_{\mathrm{o}} \qquad (8.1)$$

If the concentrations are shifted so that [A] increases by a small amount Δx (and [B] likewise diminishes) the return to equilibrium is given by

$$-\frac{\mathrm{d}[\mathrm{A}]}{\mathrm{d}t} = -\frac{\mathrm{d}\,([\mathrm{A}]_{\mathrm{o}} + \Delta x)}{\mathrm{d}t} = k_1\,([\mathrm{A}]_{\mathrm{o}} + \Delta x) - k_2\,([\mathrm{B}]_{\mathrm{o}} - \Delta x)$$

or

$$-\frac{\mathrm{d}\,\Delta x}{\mathrm{d}t} = k_1\,[\mathrm{A}]_{\mathrm{o}} - k_2\,[\mathrm{B}]_{\mathrm{o}} + k_1\,\Delta x + k_2\,\Delta x.$$

Combining with (8.1), reduces this to $-\dfrac{\mathrm{d}\,\Delta x}{\mathrm{d}t} = (k_1 + k_2)\,\Delta x.$ (8.2)

Thus the approach to equilibrium is described by a simple first-order rate law and may be characterized by the relaxation time $\tau = 1/(k_1 + k_2)$.

Let us now consider what happens when the system becomes somewhat more complex, for example

$$\mathrm{A} \underset{k_2}{\overset{k_1}{\rightleftharpoons}} \mathrm{B} + \mathrm{C}$$

At equilibrium

$$-\frac{\mathrm{d}[\mathrm{A}]}{\mathrm{d}t} = 0 = k_1[\mathrm{A}]_o - k_2[\mathrm{B}]_o[\mathrm{C}]_o.$$

Displacing the system by a small increment Δx as before gives

$$-\frac{\mathrm{d}[\mathrm{A}]}{\mathrm{d}t} = -\frac{\mathrm{d}([\mathrm{A}]_o + \Delta x)}{\mathrm{d}t} = k_1([\mathrm{A}]_o + \Delta x) - k_2([\mathrm{B}]_o - \Delta x)([\mathrm{C}]_o - \Delta x)$$

which simplifies to

$$-\frac{\mathrm{d}\,\Delta x}{\mathrm{d}t} = k_1\Delta x + k_2[\mathrm{B}]_o\,\Delta x + k_2[\mathrm{C}]_o\,\Delta x - k_2(\Delta x)^2.$$

This expression no longer reduces to linear form unless the second order term in $(\Delta x)^2$ can be neglected, in which case it becomes

$$-\frac{\mathrm{d}\,\Delta x}{\mathrm{d}t} = (k_1 + k_2[\mathrm{B}]_o + k_2[\mathrm{C}]_o)\,\Delta x$$

and

$$\tau = \frac{1}{k_1 + k_2([\mathrm{B}]_o + [\mathrm{C}]_o)}.$$

The approximation introduced here of neglecting the second-order term is fundamental to all relaxation techniques. Provided the displacement is small enough to ensure the validity of this *linearization* procedure, the approach to equilibrium is truly exponential and is characterized by a single relaxation time, which may be a complex function of the rate constants and concentrations involved.

Let us now turn to a system which involves two consecutive equilibria, for which the simplest example is

$$\mathrm{A} \underset{k_2}{\overset{k_1}{\rightleftharpoons}} \mathrm{B} \underset{k_4}{\overset{k_3}{\rightleftharpoons}} \mathrm{C}.$$

The complete solution is rather cumbersome and occupies more space than is available here, but the results may be appreciated by considering the particular case in which the first equilibrium is established more rapidly than the second. If all three concentrations are perturbed A and B will relax to a partial equilibrium while the concentration of C remains largely unchanged at its perturbed value. This first relaxation is characterized by a relaxation time τ_1 of $1/(k_1 + k_2)$ as before.

The second stage in the relaxation may be treated as follows: the concentration C is increased by a small amount Δx and A and B are reduced by small amounts Δa and Δb which together must equal Δx. Before significant change occurs in C, A and B readjust to the new partial equilibrium defined by

$$\frac{[\mathrm{B}]_\mathrm{o} - \Delta b}{[\mathrm{A}]_\mathrm{o} - \Delta a} = \frac{k_1}{k_2}.$$

Since $\Delta a + \Delta b = \Delta x$, simple algebra shows that

$$\Delta b = \frac{k_1}{k_1 + k_2} \cdot \Delta x.$$

The kinetic equation for the relaxation of C may be written

$$-\frac{\mathrm{d}[\mathrm{C}]}{\mathrm{d}t} = -\frac{\mathrm{d}([\mathrm{C}]_\mathrm{o} + \Delta x)}{\mathrm{d}t} = -\frac{\mathrm{d}\Delta x}{\mathrm{d}t} = k_4[\mathrm{C}] - k_3[\mathrm{B}]$$

$$= k_4([\mathrm{C}]_\mathrm{o} + \Delta x) - k_3\left([\mathrm{B}_\mathrm{o}] - \frac{k_1}{k_1 + k_2} \cdot \Delta x\right).$$

Subtracting the equilibrium condition $k_4[\mathrm{C}]_\mathrm{o} = k_3[\mathrm{B}]_\mathrm{o}$ as before, gives once again a simple first-order relation

$$-\frac{\mathrm{d}\Delta x}{\mathrm{d}t} = \left(k_4 + \frac{k_3 \quad k_1}{k_1 + k_2}\right) \Delta x$$

and the second relaxation time τ_2 is $1/\{k_4 + k_3 k_1/(k_1 + k_2)\}$.

This exercise demonstrates that the relaxation of a two-step equilibrium is characterized by two separate relaxation times. These relaxation equations are strictly linear and do not contain concentration terms because only first-order processes are involved.

The sole assumption made was that the two equilibria are established at significantly different rates; a more general analysis, without this assumption, produces similar results but with more involved expressions for τ_1 and τ_2.

For a general multistep system, appropriately linearized if higher order processes occur, there exists a *relaxation spectrum*, comprising a series of relaxation times corresponding to the number of steps involved. *Relaxation spectroscopy* is concerned with the resolution of a complex relaxation process into its component relaxation times.

As so often in kinetics, there are two separate ways in which relaxation studies are undertaken. In the most obvious, a perturbation, on this occasion quite small, is applied to the system and the subsequent change in some property of the system is followed directly in real-time. In the alternative approach, discussed later, a sinusoidal perturbation is applied so that a steady-state situation is established. The three physical properties most readily perturbed are temperature, pressure, and electric field gradient.

The temperature-jump method

Perhaps the best-known of the direct relaxation techniques is the *temperature-* or *T-jump* method. The principle here is that the position of equilibrium is disturbed by the application of a very fast temperature pulse, the subsequent relaxation being followed by monitoring another physical property of the system. The technique is in principle applicable to any reaction which involves a significant enthalpy change, ΔH° (remember† that the equilibrium constant of a reaction has a temperature dependence governed by $\Delta H/RT^2$).

In ionic systems the temperature rise is achieved by discharging a condenser through the solution in much the same way as in a flash photolysis discharge. The ohmic heating raises the temperature by a few degrees within a period of about a microsecond. The ions merely serve to carry the current and it is not essential that the chemical reactions involve these ions. Most systems investigated in this way have indeed been ionic and in such cases it is frequently possible to use the conductivity of the solution as a measure of the extent of reaction. Alternatively a spectrophotometric method may be employed, possibly with the addition of an indicator if the system does not itself contain a suitable chromophore. It should be emphasized that the aim is simply to monitor the departure of the system from equilibrium and not to observe transient species as would normally be the case with large perturbation techniques. Although high sensitivity is required, because the perturbation is minimal and the cell volume is small, the detection technique does not have to be particularly selective.

A typical T-jump apparatus is shown in Fig. 8.1. A potential of 100 kV is employed in conjunction with a 0.02 μF capacitance and can generate a temperature rise of 6° within a tenth of a microsecond in a volume of

† See Smith's *Basic Chemical Thermodynamics* (OCS 8) for the thermodynamic description of the dependence of K on pressure, volume, and temperature.

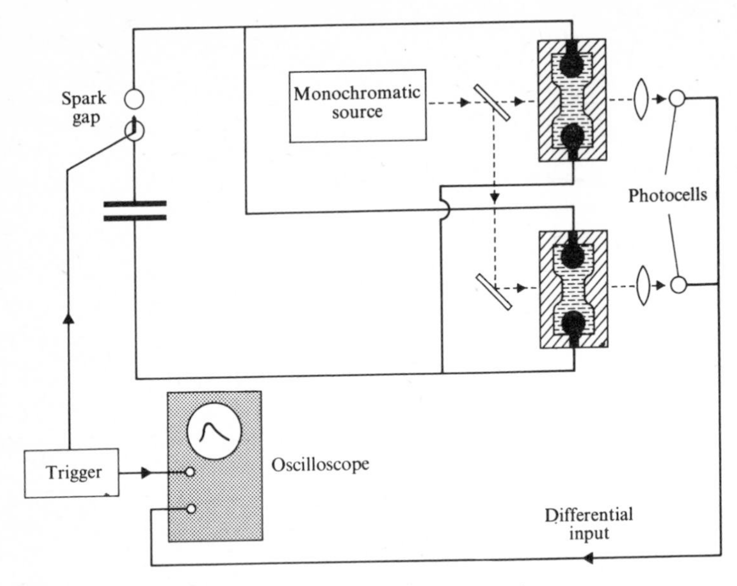

Fig. 8.1. Temperature-jump apparatus.

1 ml. It is desirable to employ a pair of matching cells, which form two arms of a Wheatstone net, if the conductivity is to be monitored. For spectrophotometric measurements, a monochromatic beam of light is passed through both sample and reference cells and the difference in transmission is displayed on an oscilloscope. Uniform heating of the liquid is essential because temperature gradients would distort the optical path due to changes in the refractive index.

For non-conducting liquids, a similar temperature pulse may be obtained either from a microwave source, applying the high potential pulse to a suitable magnetron, or alternatively by using a Q-switched laser. The problem with lasers is the difficulty of transforming the photon energy into temperature energy within short time intervals because the common solvents are transparent to the output of the usual solid-state lasers. The addition of a small quantity of an inert dye assists the conversion of the laser power but the process is too inefficient for use with Q-switched lasers.

The temperature-jump method has been used extensively to study one of the most fundamental of chemical reactions: the neutralization of protons by hydroxyl ions

$$H^+ (H_3O^+) + OH^- \rightarrow H_2O.$$

The temperature change produces an equivalent change in the conductivity and the subsequent relaxation occurs with $\tau \sim 40\,\mu s$ at room temperature, corresponding to a bimolecular rate constant of 1.3×10^{11} $\ell\,mol^{-1}\,s^{-1}$. An interesting feature of this rate constant is that it is higher than permitted by simple diffusion considerations since it proves necessary to postulate a distance of closest approach of 0.75 nm (7.5 Å). This somewhat unexpected finding has been interpreted in terms of the concerted transfer of protons along hydrogen bonds illustrated in Fig. 8.2.

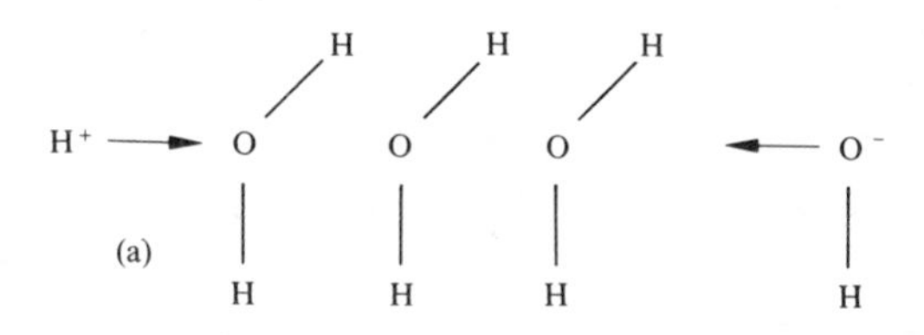

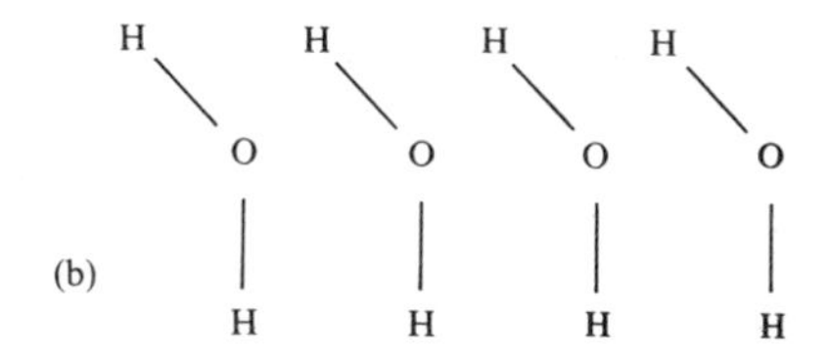

Fig. 8.2. Neutralization by concerted proton transfer.

The T-jump method has been applied to a wide range of related proton and hydroxyl ion transfer reactions, all of which seem to occur at the diffusion-controlled limit unless prevented by special circumstances such as intramolecular hydrogen bonding or solvent reorganisation. It has also been used to study the formation of metal complexes, which usually occurs via outer-sphere complexes

$$M{\cdot}H_2O + L \rightleftharpoons M.H_2O.L \rightleftharpoons M.L\;(+\,H_2O).$$

The rate-determining step is the loss of the water molecule from the intermediate.

In recent years, the temperature-jump method has become the principal tool for studying the enzyme reactions described in Chapter 7;

perhaps the major contribution has been the confirmation of reaction schemes of greater complexity than envisaged by the original Michaelis-Menten concept.

The electric-field displacement method

An applied electric field will influence equilibria involving ions, dipoles, or polarizable species. The magnitude of the field required is high, typically hundreds of kilovolts per centimetre, in liquids of high dielectric constant such as water. An electric field acts upon a solution of a weak electrolyte by increasing the ionic conductivity (*first Wien effect*), because of the distorted ionic atmosphere, and by increasing the degree of dissociation (*second Wien effect*). It is the latter which is of importance and the former must therefore be kept to a minimum.

The experimental arrangement is virtually identical to that employed for temperature jump measurements, the output from a charged condenser being applied to the electrodes in a similar cell. In this case, the resistance of the cell is kept high so that little current is drawn. This minimizes ohmic heating and helps to maintain a uniform field for the required period.

The associated electronic circuitry proves more complex, mainly because the square-pulse must be of short duration ($< 10^{-4}$ s) if secondary space charge and polarization effects are to be avoided. Although it would appear possible to short-circuit the capacitor at a short interval after its application to the electrodes, this proves difficult to achieve. The most common way of overcoming the problem is to discharge a *delay line* or co-axial cable of appropriate length into a matching impedance: the duration of the square pulse is simply the time taken to travel the length of the cable at the velocity of light.

When shorter times are required, say 10^{-6} s or less, the square pulse may be replaced by a damped harmonic pulse. A series of experiments is conducted with different periods and the maximum response measured. Although only a single pulse is employed, the technique really lies closer to the category of periodic relaxation methods dealt with in the next chapter.

In general, the electric-field displacement method is well-suited to the study of weak electrolytes. Relaxation times down to 10^{-7} s, significantly shorter than available by the T-jump method, can be measured although problems arise with times in excess of 10^{-4} s. The technique has been used in the study of proton transfer processes and indeed the $H^+ + OH^-$ reaction rate was first obtained in this way. It has also been applied to the hydrolysis of trivalent metal ions e.g.

$$M^{3+} + H_2O \rightleftharpoons MOH^{2+} + H^+$$

$$MOH^{2+} + H_2O \rightleftharpoons M(OH)_2^+ + H^+$$

with M = Al, Sc, Cr, In, and Ga; although two relaxation times were expected, only one could be detected.

The pressure-jump method

If a reaction involves a change in volume, as in many ionic systems, the position of equilibrium depends on the pressure.† It is therefore possible to create the necessary perturbation by rapidly changing the pressure exerted on the system.

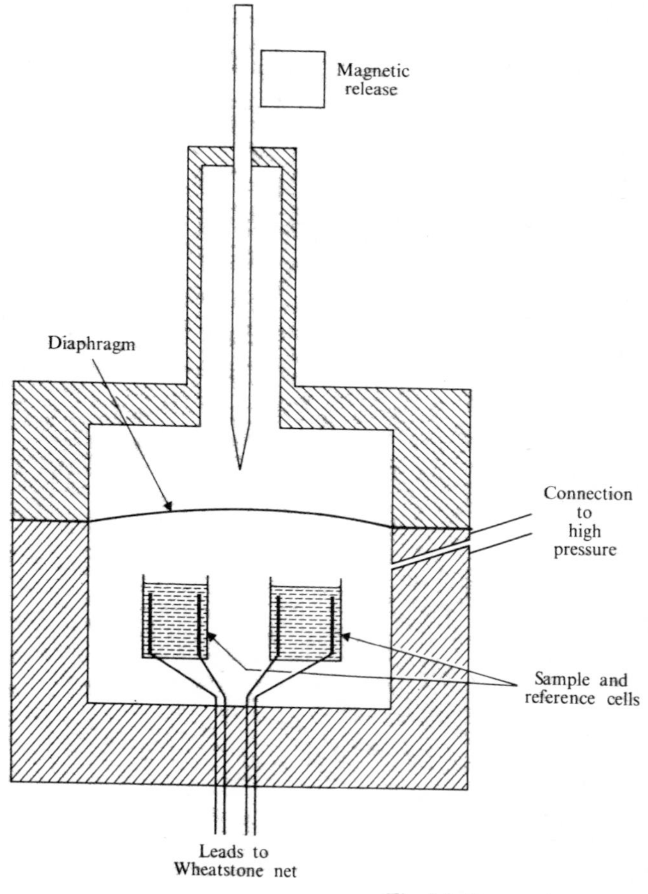

Fig. 8.3. Pressure-jump apparatus.

† Once again, see Smith, *loc. cit.*

The apparatus which has been used for the purpose is illustrated in Fig. 8.3. The reaction cell is pressurized up to *ca.* 5 MN m^{-2} (50 atm) by admitting compressed gas from a commercial cylinder through a fine capillary. The pressure is then released by rupturing a thin metal diaphragm in much the same way as is used to generate a shock wave. The capillary serves to prevent drastic loss of gas should the diaphragm break prematurely. Because the expansion process occurs via an *expansion*, or *rarefaction*, *wave*, the pressure change is rather slower than the corresponding shock transition taking about 10^{-4} s.

The alternative, shock transition has also been used to create a pressure jump (Fig. 8.4). This causes a greater pressure change (up to one GN m^{-2}

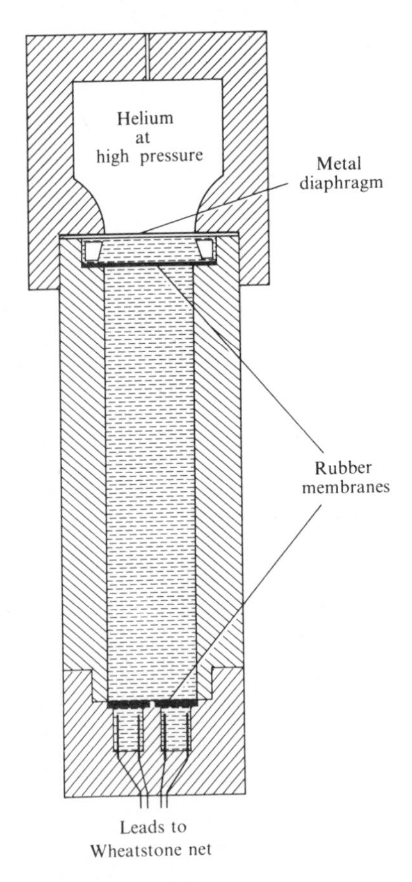

Fig. 8.4. Shock tube for liquids.

or 10^4 atm) in a shorter time interval (less than 10^{-8} s) but the relaxation can be studied only for short periods (up to 10^{-4} s) because of subsequent wave phenomena. There is effectively no upper limit to the relaxation time which can be followed by the rarefaction technique.

The relaxation has normally been followed by monitoring the change in conductivity of the liquid, although other techniques can be employed. The conductivity is affected by solvent volume and ion mobility changes in the liquid as well as by the change in the position of equilibrium so that it is preferable to incorporate within the pressure chamber a reference cell containing the same solvent plus a non-reacting electrolyte.

Pressure-jump measurements have been made on the formation of ionic complexes e.g. the reactions of ferric ions with various anions, where ΔV° is appreciable.

It is fair to say that the pressure-jump method has proved much less valuable than the temperature-jump method. This is because the effect produced by a 10° temperature change is normally much greater than that produced by a 50 atmosphere pressure change. Furthermore, in the absence of reaction, a temperature change does not interfere with the optical detection technique in the way that conductivity is affected by a pressure change.

9. Periodic relaxation methods

THE previous chapter dealt with the relaxation of an equilibrium system which has been subjected to a single perturbation; the experimental techniques introduced below depend on the use of a fluctuating perturbation. The principle on which such techniques operate may be appreciated by employing the concept introduced previously of a characteristic relaxation time. If the perturbation is applied sufficiently slowly, then the chemical equilibrium will 'follow' the disturbance throughout the cycle. On the other hand, if the perturbation oscillates very rapidly, with a period much less than the relaxation time, the equilibrium will be unable to adjust to the disturbance before it has reversed. The net effect is that the position of equilibrium becomes virtually 'frozen'. In order to measure the relaxation time it is necessary first to discover some property of the system which varies depending on whether the equilibrium is 'active' or 'inactive' and then to measure the frequency of the oscillating perturbation at which transition between the two states occurs. The major advantage of this approach is that it utilizes the principle of competition so that a stationary state is established and the problem of working in real-time is avoided.

The basic relaxation equation (eqn 8.2) described in the previous chapter may be written in the form

$$-\frac{\mathrm{d}x}{\mathrm{d}t} = \frac{x - x_0}{\tau} \qquad (9.1)$$

where x_0 corresponds to the final equilibrium situation. x_0 must now be replaced by a sinusoidally-varying quantity, $x_0 = A\,\mathrm{e}^{i\omega t}$, which is termed the *forcing function.* (It would be more correct to use a sine or cosine function, but the exponential function proves mathematically easier to handle and leads to the same conclusion.) After the stationary state has been established, x must show the same frequency dependence as x_0 and hence takes the form

$$x = B\mathrm{e}^{i\omega t}$$

from which

$$\frac{\mathrm{d}x}{\mathrm{d}t} = i\omega B\mathrm{e}^{i\omega t} = i\omega x.$$

Substitution in (9.1) gives

$$i\omega x = \frac{x - x_0}{\tau} \quad \text{or} \quad x = x_0\left(\frac{1}{1 - i\omega\tau}\right).$$

The quantity $\dfrac{1}{1 - \mathrm{i}\omega\tau}$ is termed the *transfer function* and may be separated into real and imaginary parts, thus

$$\chi = \chi_{\text{real}} + \chi_{\text{imaginary}} = \left(\frac{1}{1+\omega^2\tau^2}\right) + \mathrm{i}\left(\frac{-\omega\tau}{1+\omega^2\tau^2}\right).$$

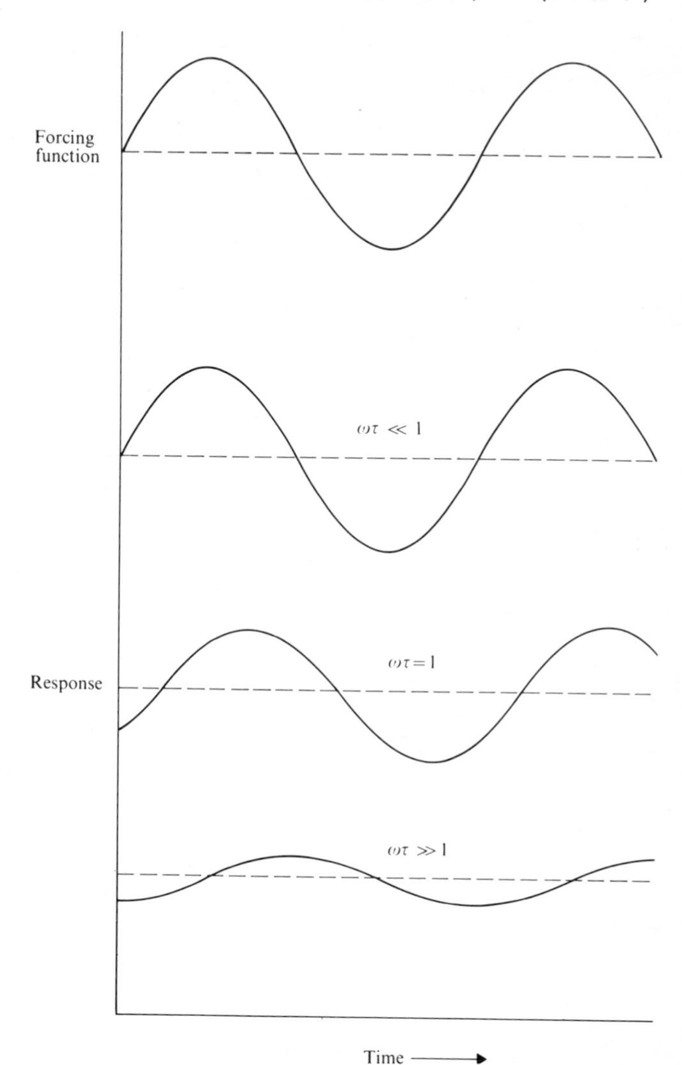

Fig. 9.1. Possible responses to a sinusoidal forcing function.

$\omega\tau$, the product of the angular frequency and the relaxation time, is thus a dimensionless quantity which describes the response of the system to the forcing function.

The significance of the separation of the transfer function into real and imaginary parts is that the former denotes the component of the response which remains in phase with the forcing function while the latter denotes the component which is 90° out-of-phase. The response to a sinusoidal function is shown in Fig. 9.1 and it can be seen that when $\omega\tau$ is very small ($\chi_{real} \gg \chi_{imaginary}$) the response is almost exactly in phase with the perturbation whilst when $\omega\tau$ is very large ($\chi_{imaginary} \gg \chi_{real}$) the response is almost 90° out of phase and is very small. The two components are equal when $\omega\tau = 1$ and the response is then exactly 45° out-of-phase. The plot of the real and imaginary components in Fig. 9.2 shows that a measurement of either can be used to locate the

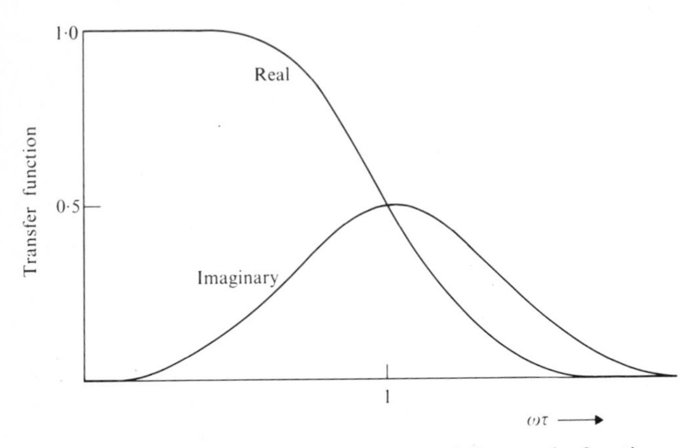

Fig. 9.2. Real and imaginary components of the transfer function.

position of $\omega\tau = 1$ and indeed two quite distinct physical measurements are available for the purpose.

A typical physical property which shows a frequency dependence corresponding to the real part of the transfer function is the sound velocity. The dependence of velocity on frequency is called *dispersion* and so methods which are related to the real part are termed *dispersion* methods. The imaginary part, which measures the out-of-phase character of the response, represents the energy loss to the system due to relaxation and hence methods which employ this component are termed *absorption* methods. It will be noted that optical properties display similar characteristics and employ the same terminology. This treatment is quite general and applies equally well to the effect of sound waves on pressure- and temperature-dependent equilibria (ultrasonic relaxation), or of alternating electric fields on systems containing charged or polar species (dielectric relaxation).

Ultrasonics

The study of sound propagation demands devices for creating sound waves and subsequently for their detection. Both functions are achieved by means of piezo-electric crystals; such crystals generate an electric signal when subjected to mechanical stress or, conversely, change their shape when an electric field is applied. The most common material in use is quartz and the resonant frequency is controlled by cutting the crystal to the appropriate dimensions. The sound wave is generated by applying an oscillating electric field of that frequency. At high frequencies, where the dimensions of the crystal must be small, it is preferable to drive a crystal at a frequency other than its resonant frequency.

Many different experimental configurations have been adopted for ultrasonic studies, of which only a selection can be presented here.

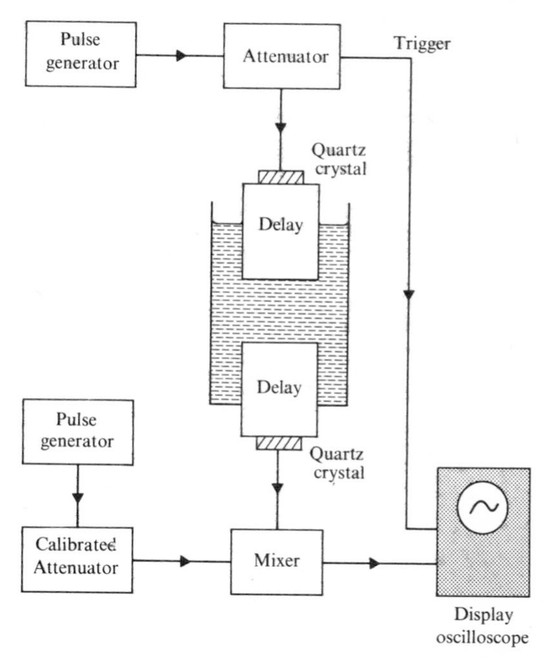

Fig. 9.3. Pulse technique for ultrasonic wave propagation studies.

The pulse technique

This is the most common method of studying the propagation of ultrasonic waves in liquids and permits the measurement of very short relaxation times (Fig. 9.3). A radio-frequency oscillator is arranged to give an alternating signal of the desired frequency and the output passes through a *gating* circuit which transmits r.f. pulses at regular intervals. This excitation voltage is applied to a quartz transducer which generates the pulse of ultrasonic 'radiation' at the upper end of a fused quartz block known as a *delay line.* The pulse travels down the delay line, through the medium under investigation, and into a second delay line at the base of which the pulse is detected by a second transducer. The function of the delay lines is simply to ensure that stray electrical pick-up from the original pulse has died away by the time the transmitted pulse arrives at the detector. The amplitude of the detected pulse is measured by comparing it with a reference r.f. pulse that has passed through a calibrated attenuator. The sound absorption is then obtained by following the signal intensity as a function of distance of travel through the medium, and the time the pulse takes to traverse the sample provides a measure of the sound velocity.

The ultrasonic interferometer

The ultrasonic interferometer is more often used to study relaxation in gases but can also be employed with liquids. The principle here is that signals from a resonating oscillator impinge on a plane reflector and travel back to the quartz transducer (Fig. 9.4). When the distance

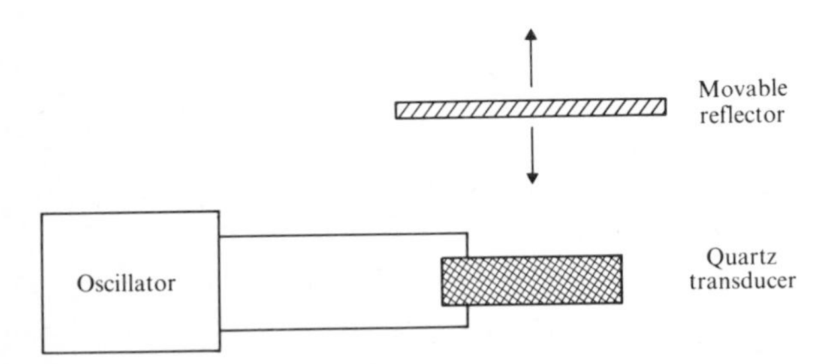

Fig. 9.4. Principle of ultrasonic interferometer.

between reflector and transducer is an integral number of half-wave-lengths, the signal returned to the transducer is 180° out-of-phase with the vibrating crystal due to the phase shift on reflection. This reduces the amplitude of the crystal oscillations and hence the current drawn by the crystal. The technique is basically simple although strict pre-cautions regarding parallelism, the avoidance of reflections from the container walls, etc., are needed to ensure meaningful results. Another serious disadvantage is that the crystal has to be operated at its natural resonant frequency. The technique can produce accurate velocity measurements but is less suited to the study of absorption.

The reverberation method

The liquid under investigation is contained in a thin glass sphere, typically of 5 litre capacity, and is excited by a transducer to give radial vibrations at its resonant frequency (Fig. 9.5). The driving transducer is

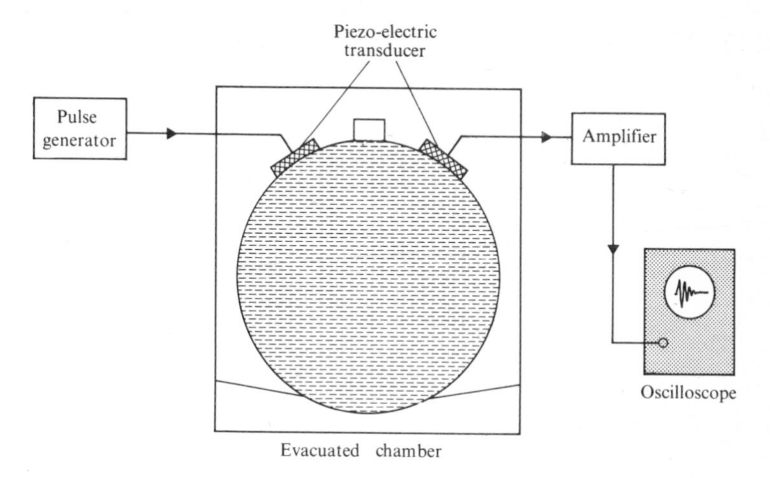

Fig. 9.5. The reverberation method.

then switched off and a second transducer is used to monitor the radial vibrations as they decay, the rate depending on the absorption coefficient of the liquid. The technique cannot be employed for absolute measurements but is very useful for comparative measurements in dilute solutions as the pure solvent may be used for calibration purposes.

Energy transfer in gases

Acoustic methods have not yet been developed for the study of chemical reactions in gases although they have proved invaluable for investigating the faster energy transfer processes which are so important to an understanding of gas-phase kinetics. Energy is distributed among translational, rotational, vibrational, and electronic modes in molecules, and chemical reactions, which involve the making and breaking of chemical bonds, depend critically on the exchange of energy between the translational and vibrational modes. For all but the lightest molecules, transfer of energy between translation and rotation is efficient so that these two motions can normally be regarded as fully equilibrated throughout the course of reaction. However, rotational motions do appear to influence the efficiency of energy transfer between translation and vibration and must therefore be taken into consideration. Most of our knowledge of vibrational energy transfer comes from acoustic studies although such techniques are unsuited to high temperature and above 900 K the shock tube has provided the main source of information.

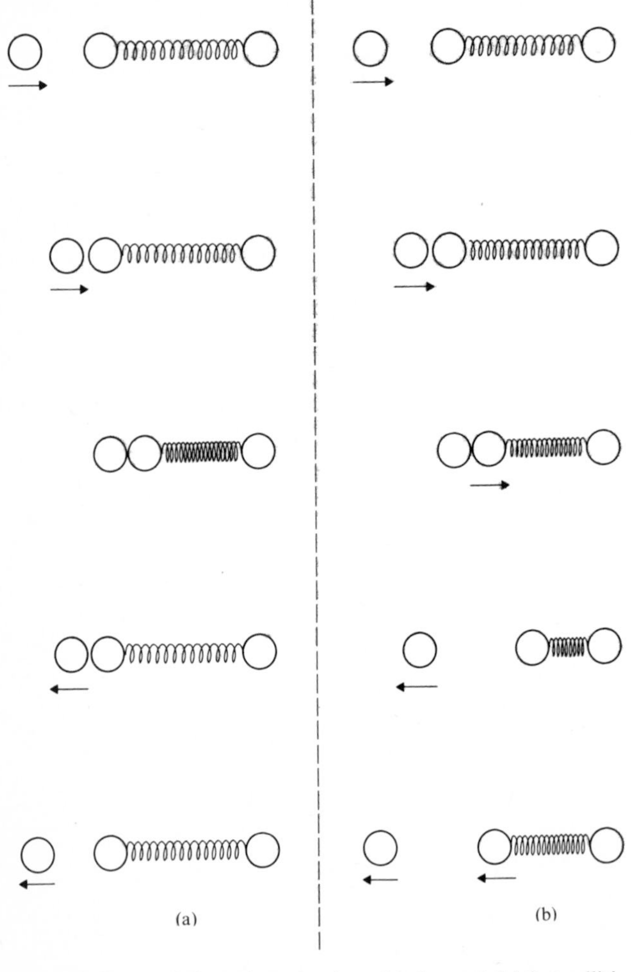

Fig. 9.6. Response of diatomic molecule to (a) slow and (b) fast collisions.

Although the detailed theory is complex we can understand something about the process from the following model. Suppose that a diatomic molecule can be represented by a pair of spongy spheres connected by a simple spring. It has already been shown that, if the relaxation time is short compared with the period of the forcing function, the system will remain in equilibrium throughout the perturbation. In the normal collision of an atom with a diatomic molecule, the velocity of approach is about 10^3 m s^{-1} and the range of the interaction exceeds 10^{-10} m so that the duration of the collision is greater than 10^{-13} s. A typical vibration frequency of 3000 cm^{-1} (100 THz) has a characteristic period of 10^{-14} s, so that the interatomic distance in the molecule is able to adjust to the perturbing collision, returning subsequently to its original value (Fig. 9.6). If the collision is of shorter duration, the interatomic separation cannot adjust and the collision partner leaves the diatomic species in a perturbed state with the atoms moving towards each other. The exchange of energy is thus an infrequent process involving only molecules moving with higher-than-average velocities. The efficiency of energy transfer is however expected to increase with increasing temperature (due to the higher fraction of fast-moving molecules), lower vibration frequency, and steeper interaction potential (implying a shorter range interaction or 'harder' molecules). These effects, first put into quantitative form by Landau and Teller and subsequently modified by Schwartz, Slawsky, and Herzfeld, are all borne out by ultrasonic measurements: molecules like nitrogen and carbon monoxide with high vibrational frequencies exchange energy only once in every 10^{10} collisions at room temperature while simple organic molecules like propane and methanol relax in less than ten.

An interesting feature of the energy transfer process is that two molecules readily exchange energy if they possess motions which are nearly resonant. Nitrogen, with a vibration frequency of 69.9 THz (2330 cm^{-1}), exchanges energy rapidly with carbon dioxide, which has a frequency of 70.4 THz (2346 cm^{-1}). Since carbon dioxide has a shorter T $\leftrightarrow$ V relaxation time than nitrogen, a trace of the former drastically reduces the effective T $\leftrightarrow$ V time for the latter.

Reactions in solution

Ultrasonic techniques have proved very powerful for the study of reactions in solution, largely because half-lives down to 10^{-10} s may be followed. Energy transfer can be investigated in liquids as well as in gases and proton transfer reactions, similar to those described in the previous chapter, have been examined e.g.

$$H_3O^+ + SO_4^{2-} \rightleftharpoons HSO_4^- + H_2O$$

$$H_3O^+ + SO_3^{2-} \rightleftharpoons HSO_3^- + H_2O$$

$$NH_4^+ + OH^- \rightleftharpoons NH_4OH.$$

Other applications have been in the study of rotational isomerizations and conformational changes.

One reaction which has been quite extensively studied is the rate of association, or hydrogen bond formation, in carboxylic acids. The process can be quite complex, in the case of acetic acid displaying two separate relaxation times tentatively attributed to the formation of one and two hydrogen bonds, thus:

$$2\,CH_3{-}C({=}O){-}OH \rightleftharpoons CH_3{-}C({=}O){-}OH\cdots O{=}C(CH_3){-}HO \rightleftharpoons CH_3{-}C\begin{matrix} {=}O\cdots HO{-} \\ {-}OH\cdots O{=} \end{matrix}C{-}CH_3.$$

Perhaps the most interesting results have come from the study of ion-pair formation. Acoustic measurements on 2:2 electrolytes in the 0.01 to 0.1 M range show two distinct absorption maxima corresponding to two separate relaxation times. Neither of these can be attributed to the actual association process and it appears that they correspond to the loss of solvated water molecules from the solvent-separated ion pair, thus

$$[(M^{2+}H_2O)(H_2O\,A^{2-})] \rightleftharpoons [(M^{2+}H_2O)\,A^{2-}] \rightleftharpoons [M^{2+}\,A^{2-}]$$

Loss of water occurs first from the anion with a rate constant of about $10^9\,s^{-1}$: loss from the cation then follows with a rate constant in the range 10^2 - $10^7\,s^{-1}$ depending on the nature of the cation.

Dielectric relaxation

The capacitance *in vacuo* C of a parallel-plate condenser is given by q/V where q is the charge on the plates and V is the applied potential (Fig. 9.7). If a material is inserted between the plates, it is *polarized* by the field, the surfaces acquiring a charge δq in the opposite sense and the charge on the plates increasing correspondingly. The new capacitance C' is equal to $(q + \delta q)V$ and the *permittivity* ϵ of the material is defined as $C'/C = (q + \delta q)/q$. The local electric field within the material is thus reduced, relative to that in vacuo, by the polarization.

The polarization arises from two main sources: *distortion polarization*, where the molecular structure is deformed by the field, either by shifting the electrons with respect to the nuclei or the atoms with

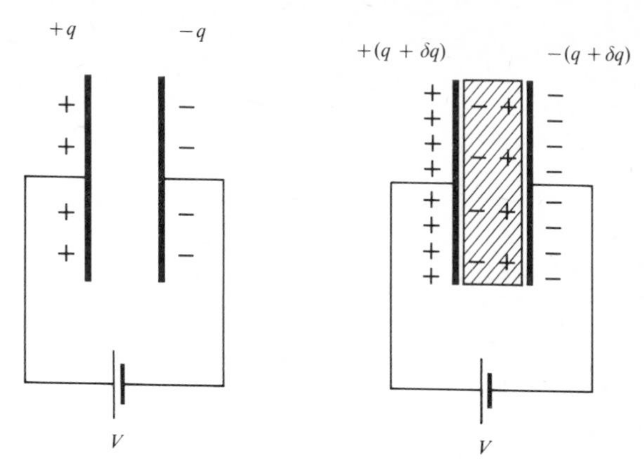

Fig. 9.7. The effect of polarization on the capacitance of a condenser.

respect to each other, and *orientation polarization*, where the molecules with permanent dipoles tend to align themselves parallel to the direction of the applied field. The total orientation effect is small because it is opposed by ordinary thermal motions which attempt to preserve the random distribution.

If the electric field is varied sinusoidally rather than remaining steady the polarization will keep in phase with the field. However as the frequency is raised a stage is eventually reached at which the polar molecules cannot rotate sufficiently rapidly. The contribution of the orientation polarization to the total polarization is thus lost and the permittivity falls.

The overall permittivity may be divided into two parts $\epsilon = \epsilon' - i\epsilon''$ where ϵ' is the *real permittivity* (*static permittivity*, or *dielectric constant*, in the low frequency limit) and ϵ'', the imaginary part, is the *dielectric loss factor.* The latter represents the energy which is absorbed by the material. The *loss angle* δ which indicates the lag between the variation of the field and the re-adjustment of the molecules is given by $\tan\delta = \epsilon''/\epsilon'$. The values of ϵ' and ϵ'' vary with frequency in the manner shown in Fig. 9.2. A more common representation is the Cole-Cole plot of ϵ'' against ϵ' (Fig. 9.8). For simple liquids, this plot is a semicircle but for more complex fluids the curve falls inside the semicircle. This is attributed to the existence of a symmetrical distribution of relaxation times rather than one single value.

There are a number of methods available for determining ϵ' and ϵ''. Since the suitability of a particular technique depends on the frequency required, two methods have been selected which are used in the range, 1 MHz to 5 GHz.

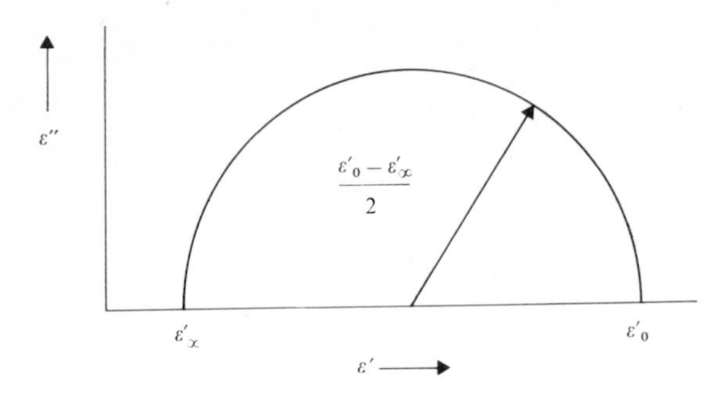

Fig. 9.8. Cole-Cole plot.

The resonant circuit method

The Hartshorn-Ward method is illustrated in Fig. 9.9. On the left is an oscillator circuit whose frequency can be adjusted as required: the right-hand section loosely coupled to it is the resonant circuit. The permittivity of a sample is obtained from the relation $\epsilon' = d/d_0$, where d and d_0 are the separations of the parallel plates of the variable condenser, with and without the sample, at which resonance occurs.

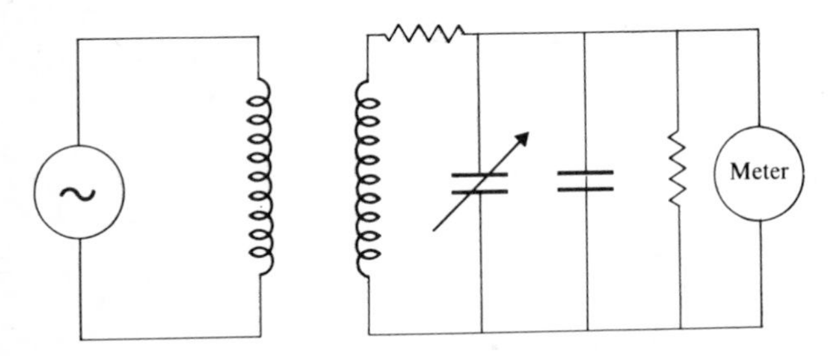

Fig. 9.9. Circuit for Hartshorn-Ward method.

Although the signal at the maximum is determined by ϵ', at positions away from resonance it depends on ϵ''. The method is therefore applicable to both dispersion and absorption determinations although the accuracy of measurement of ϵ' is better than that of ϵ'' (0.1 per cent *cf.* 1 per cent).

The transmission-line method

This method is based on the wave-propagation properties of transmission lines (Fig. 9.10). The transmission lines used may comprise either the region between two coaxial conductors i.e. concentric cylinders (100 MHz to 5 GHz) or else simply hollow waveguides (3 GHz

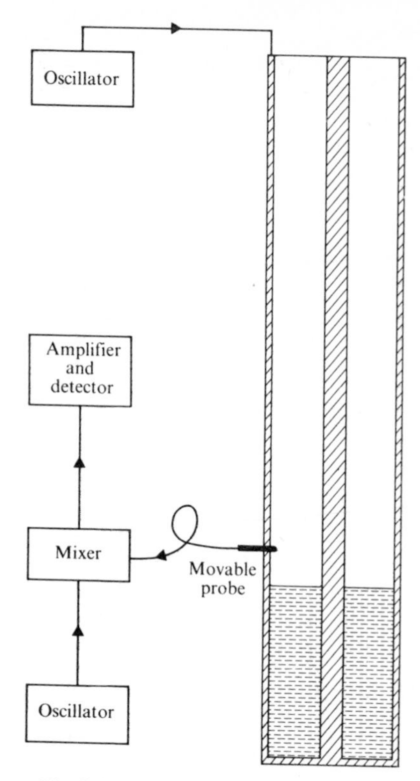

Fig. 9.10. Transmission-line technique.

to 60 GHz). An oscillating signal is applied at one end and the transverse wave which propagates down the line reflects at the closed end so that a standing-wave pattern becomes established. The position of the nodes and the ratio of the maxima to the minima are obtained from a sliding probe inserted through a narrow slot in the outside wall of the container. The mathematics involved in obtaining ϵ' and ϵ'' from the experiments is complex and will not be described here: it is worth noting that the calculations are considerably simplified if the sample depth is adjusted exactly to one-quarter of a wavelength. The accuracy of the technique is comparable with that of the previous method.

Applications

Dielectric relaxation is used to study rates of rotation, or rotational relaxation times, of polar molecules in the liquid and solid phases. In the narrowest sense of the term 'reaction', the technique has not proved particularly valuable. However, for dynamic processes in general, the quantity of information it has provided on conformational changes in molecules, flexing of polymers, and helix-coil transitions has been quite outstanding.

In water, for example, dielectric relaxation yields a relaxation time of 10^{-11} s while in ice the process is much slower with $\tau \sim 10^{-5}$ s. In addition the activation energy of the process is greater in ice (55.9 kJ mol^{-1}) compared with water (21.0 kJ mol^{-1}). This implies that three hydrogen bonds must be broken in ice, but only one in water, when the molecule rotates. These measurements refer to overall rotation of the molecule but intramolecular changes, such as internal rotations, ring inversions (e.g. cyclohexanone) etc., have also been studied in a wide range of substances.

A particularly important application of dielectric relaxation has been in the study of segmental motions of polymer molecules. Most mechanical properties of solid polymers depend on the mobility of the segments, or repeating units, in the long chains and in solution the reaction rates are commonly governed by the same quantities.

10. Electrochemical techniques

ELECTROCHEMICAL methods for measuring reaction rates depend primarily on a competition between the chemical reaction and the diffusion of some species to an electrode surface. The mathematics involved tends to be rather complex even though the principles underlying the techniques are reasonably straightforward. A very much oversimplified treatment is therefore presented here and it should be appreciated that the correct relationships are far more unwieldy although formally quite similar.†

We are primarily concerned with a general electrode reaction in which a reducible species X accepts an electron at the cathode and forms the reduced species R

$$X + e^- \rightarrow R.$$

For all practical purposes the process is rapid and irreversible so that all species X which strike the electrode surface are immediately removed from the system. The current in the cell due to this reduction process is simply proportional to the rate at which the X species reach the surface.

Motion of X may be caused by three processes: *diffusion*, *convection*, and *migration*. Migration, under the influence of the electric field, is avoided by adding to the cell a large excess of *supporting electrolyte*, such as potassium chloride, which supplies the charge carriers and hence reduces the potential gradient essentially to zero without interfering with the electrode reaction. Convection occurs if the solution is stirred or temperature gradients are generated in the cell: its effect is always small in comparison with diffusion. Diffusion to the surface takes place because the species X are depleted in the vicinity of the electrode and a concentration gradient therefore builds up around it.

The concentration gradient will vary with time in the manner shown in Fig. 10.1. The *diffusion current*, which is proportional to the concentration gradient, is initially at a maximum and then falls steadily as the gradient decreases. To a good approximation, the gradient can be visualized as confined to a *diffusion layer* of thickness δ. The Einstein formula for a random walk (i.e. the drunkard's path) indicates that the displacement x varies with time t according to $x^2 = 2Dt$, where D is the diffusion coefficient. Thus the diffusion layer thickness is expected to be of the order $\sqrt{(2Dt)}$ and the flux, defined by Fick's law as $-D(\mathrm{d}c/\mathrm{d}x)$, to be

† Detailed analysis of electrochemical kinetics will be found in Albery's *Electrode kinetics* (OCS 14).

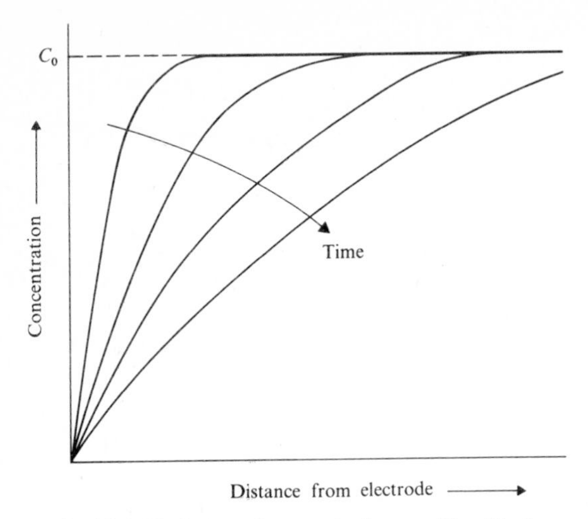

Fig. 10.1. Variation of concentration profile with time.

given by Dc_0/δ, or $2^{\frac{1}{2}}D^{\frac{1}{2}}t^{-\frac{1}{2}}$. This species flux may be converted to a diffusion current i_d by multiplying by the surface area A and the Faraday F, thus

$$i_d = 2^{\frac{1}{2}}FAD^{\frac{1}{2}}t^{-\frac{1}{2}}c_0. \qquad (10.1)$$

Let us now consider the situation which arises if X is participating in some dynamic equilibrium in the bulk solution. The simplest example may be denoted by

$$X \underset{k_b}{\overset{k_f}{\rightleftharpoons}} B$$

with the equilibrium well over to the right; the other extreme, with the equilibrium over to the left, would not affect the diffusion current. As X is depleted by reaction at the cathode and a concentration gradient is established, B will be transformed into X to replace that consumed. Thus the rate of arrival of X at the electrode surface depends both on diffusion and on chemical reaction, and with the equilibrium well to the right and the back reaction relatively slow, the current becomes completely under *kinetic control.*

The concept of a *reaction layer*, analogous to the notion of a diffusion layer, may be employed to deal with this situation. An average lifetime τ for X, which is the period of time between its formation and its removal in the dynamic equilibrium above, may also be introduced. From the same Einstein formula, species X will diffuse a distance $\mu = \sqrt{(2D\tau)}$ during this lifetime. All species X which lie less than the distance μ from the electrode surface, i.e. within the reaction layer, will be removed by reduction whilst those beyond this distance will be transformed back to B before they reach the surface. The current at the electrode is then determined by the rate at which species X are formed in the reaction layer. The manner in which the concentration gradient due to reaction is superimposed on the diffusion layer is shown in Fig. 10.2.

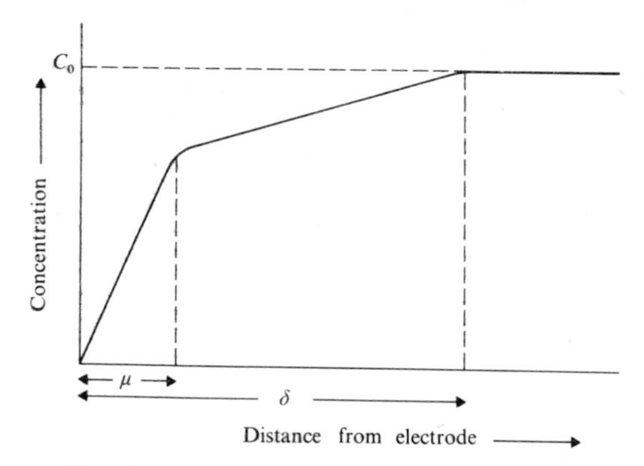

Fig. 10.2. Concepts of diffusion and reaction layers.

The volume of the reaction layer is equal to the surface area A multiplied by the thickness μ and therefore to $A\sqrt{(2D\tau)}$. If the concentration of B is c_B, the rate of formation of X within the layer is $k_b c_B A\sqrt{(2D\tau)}$. Multiplying by the Faraday as before gives the reaction current i_r as

$$i_r = F k_b c_B A\sqrt{(2D\tau)}. \qquad (10.2)$$

The mean lifetime of a transient species is given by its stationary concentration divided by its rate of disappearance (see, for example, Chapter 6 p. 60) so that τ may be equated with $1/k_f$. Since $K = k_f/k_b$, the reaction current relation becomes

$$i_r = \frac{F k_b^{\frac{1}{2}} c_B A\sqrt{(2D)}.}{K^{\frac{1}{2}}} \qquad (10.3)$$

It is conventional to express the result in terms of the ratio of the reaction current to the diffusion current which, for this elementary example, becomes

$$i_r/i_d = k_b^{\frac{1}{2}} t^{\frac{1}{2}}/K^{\frac{1}{2}}. \qquad (10.4)$$

A more rigorous treatment in which the constraint $k_f \gg k_b$ is relaxed leads to the relation

$$\frac{i_r}{i_d - i_r} = \frac{C\, k_b^{\frac{1}{2}}\, t^{\frac{1}{2}}}{K^{\frac{1}{2}}} \tag{10.5}$$

where C is a constant of order unity.

There are basically two ways in which the actual electrochemical measurement may be made. The obvious method is to hold either the current or voltage constant and follow the alteration of the other with time. The alternative is to create continuously a 'new' electrode surface

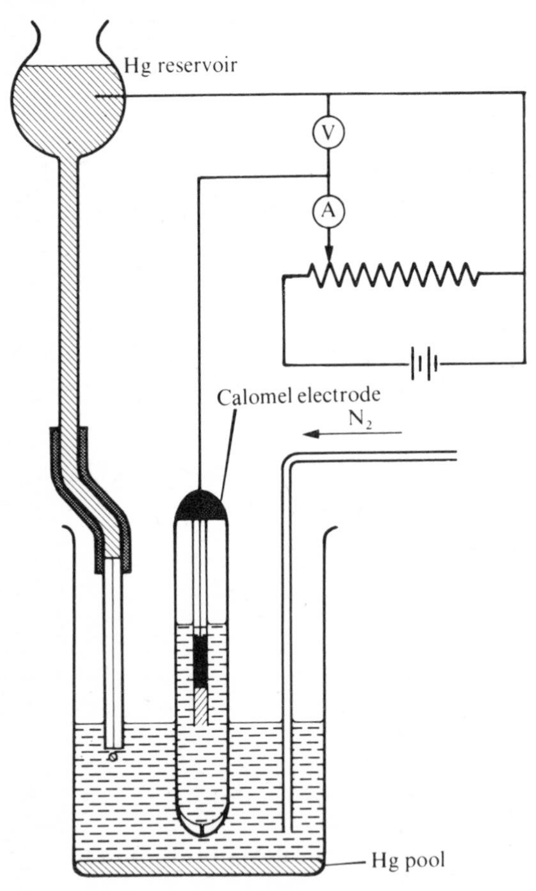

Fig. 10.3. Apparatus for polarography.

so that a steady state situation is established and a time-averaged current, very close in magnitude to the initial current, is obtained for each particular applied voltage.

Polarography

The best-known technique—*polarography*—falls in the second of the two categories. The cathode consists of a growing mercury drop at the tip of a fine capillary (see Fig. 10.3). The anode may either be a calomel electrode placed in the solution or simply the pool of mercury which forms at the bottom of the cell. As each drop leaves the capillary, a new one forms at the tip so that there is no opportunity for extensive changes in the composition of the solution even in the neighbourhood of the surface. A steadily-increasing potential is applied to the electrodes and a *polarographic wave* of the form illustrated in Fig. 10.4 is generated.† The half-wave potential $E_{\frac{1}{2}}$ is characteristic of the species

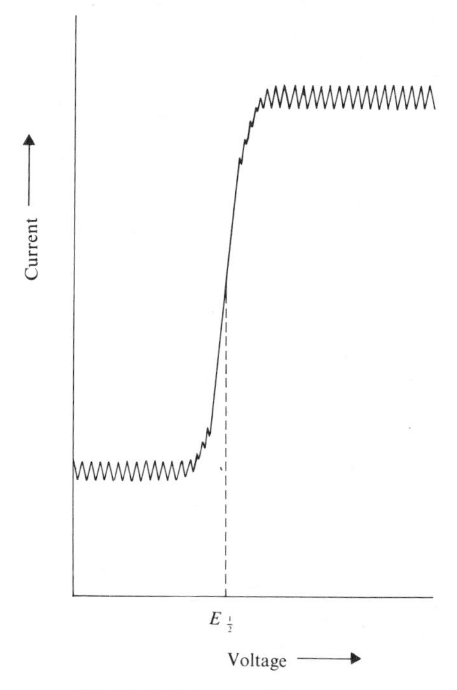

Fig. 10.4. Polarographic wave.

† See Robbins, *Ions in Solution 2* (OCS 2) for a more detailed approach, and Albery, *Electrode Kinetics* (OCS 14) for more advanced material.

being discharged and the maximum current depends on the concentration. The technique was developed originally as an analytical tool; a complete *polarogram* consists of a series of polarographic waves, each one indicating the nature and quantity of a particular ion.

The relation (10.5) between the diffusion current and the reaction current is employed for kinetic studies. The actual relationship must be modified to allow for the spherical geometry and the growth of the drop.

The rotating-disc method

As an alternative to the dropping mercury electrode, a rotating-disc may be employed to maintain constant conditions at the electrode. In practice a small cylinder of platinum is rotated about its own axis at *ca.* 5000 r.p.m., the end of the rod forming the electrode surface. The rotation causes stirring of the liquid in the immediate vicinity and thus prevents any appreciable alteration of the local conditions with time.

In contrast to the polarographic technique, the rotating disc establishes a true steady state. Close to the electrode a fixed diffusion layer exists in which the concentration gradient is constant and transport occurs at a steady rate. Outside this layer mixing serves to maintain the constant concentration. Theory not given here shows that the diffusion layer thickness δ is related to the angular velocity ω and the kinematic viscosity $\bar{\nu}$ (viscosity/density) by the relation $\delta \propto D^{\frac{1}{3}}\,\omega^{-\frac{1}{2}}\,\bar{\nu}^{-\frac{1}{6}}$.

Chronopotentiometry: constant current electrolysis

The most common time-dependent method involves maintaining a constant current and observing the voltage-time behaviour: this is sometimes termed a *galvanostatic* method.

For any given conditions the voltage shows a relatively slow change with time until it reaches the characteristic *transition* time τ_d and then it commences to rise rapidly. The cause of this behaviour may be appreciated from Fig. 10.5. As reduction at the electrode proceeds, the flux, and hence the current, will be held constant provided the concentration gradient at the electrode surface also remains constant. When the concentration actually reaches zero it is no longer possible for the gradient to remain unchanged. In order that the same current is maintained, the voltage rises rapidly until the discharge of a different species, normally the ions of the supporting electrolyte, commences.

Using a very elementary method, it is found that the diffusion current varies with time according to $i_d \propto t^{-\frac{1}{2}} c_o$ (see eqn 10.1). If the current is kept constant, it can be held at the initial value i_o only up to a time defined by $i_o t^{\frac{1}{2}} = \text{constant} \times c_o$. This is the transition time τ_d for pure

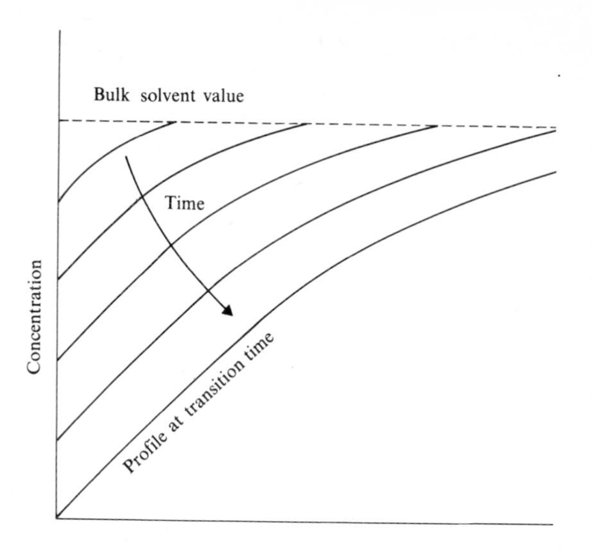

Fig. 10.5. Concentration profiles in constant current electrolysis.

diffusion. Adopting the concept of a reaction layer to deal with the situation in which the production of X is kinetically controlled as before, we may now insert the relation between current and transition time into (10.5). This yields

$$i_o \tau^{\frac{1}{2}} = i_o \tau_d^{\frac{1}{2}} - \frac{K^{\frac{1}{2}}}{C k_b^{\frac{1}{2}}} \cdot i_o$$

$$= \text{constant} - \frac{C' K^{\frac{1}{2}}}{k_b^{\frac{1}{2}}} \cdot i_o$$

Thus a plot of $i_o \tau^{\frac{1}{2}}$ versus i_o should give a straight line whose slope yields the required rate constant.

The experimental requirements for chronopotentiometry are quite minimal: the cell conveniently comprises a pool of mercury for the working cathode and a calomel electrode as the reference. The constant current can be obtained by electronic control or more simply by connecting a large resistance in series with the cell and applying several hundred volts. The voltage-time characteristics are then measured on a suitable recorder.

Applications in kinetics

These three techniques–polarography, the rotating-disc, and chronopotentiometry–as well as others not described here, provide relatively inexpensive methods for following fast reactions. To quote one example, polarography has been used quite extensively to study the dissociation and recombination of weak acids

$$HA + H_2O \rightleftharpoons A^- + H_3O^+.$$

The reduction of the hydrogen ion cannot be used as the electrode reaction because of the overvoltage on mercury. The difficulty can be circumvented by changing the nature of the electrode reaction: thus, by adding azobenzene for example, the electrode reaction becomes

$$PhN = NPh + 2H^+ + 2e^- \rightarrow PhNH - NHPh.$$

These techniques have produced rate constants varying from 10^1 to 10^{14} (!) $\ell\ mol^{-1}\ s^{-1}$ although the values at the upper end of the scale must be treated with caution. When an extremely fast reaction is being investigated, the reaction layer may well be less than 100 Å (10 nm) thick. Under such conditions, in the near vicinity of the electrode surface, the kinetic processes may not correspond with those characteristic of the bulk solution. With somewhat slower reactions, and hence greater reaction layer thicknesses, this problem should not arise.

11. Equilibrium methods

A LIMITED number of techniques permit the investigation of rate processes without disturbing the dynamic equilibria involved. Only the most important of them—nuclear magnetic resonance—will be discussed in any detail here. It will be assumed that the reader is familiar with the basic principles underlying magnetic resonance phenomena (see, for example, K. A. McLauchlan: *Magnetic resonance*, OCS 1) and with the operation of n.m.r. spectrometers, most of those in use nowadays being commercial instruments.

Nuclear magnetic resonance techniques

Spectral line broadening

The Heisenberg Uncertainty Principle states that simultaneous measurements of conjugate quantities are possible only within certain 'error' limits and gives, for energy and time, the relation $\Delta E.\Delta t \gtrsim h/2\pi$. If the average lifetime of an excited molecule is Δt, then the energy of the state will be indeterminate by the corresponding amount ΔE. If such a state radiates, the frequency emitted is governed by the Bohr relation $h\nu = E_2 - E_1$ and, as E_2 is uncertain by the amount ΔE, the emitted frequency will be correspondingly uncertain by an amount $\Delta\nu = 1/(2\pi\Delta t)$. Identical spectral transitions do not therefore occur at identical frequencies but are distributed about a mean value. The *half-width* (width at half-height) of a spectral line is thus a measure of the lifetime Δt.

The *resolving power*, $\nu/\Delta\nu$, of a high quality ultraviolet spectrometer is unlikely to exceed 10^4 (equivalent to 10^9 - 10^{11} Hz or c/s) so that true line-widths can be measured only if the life-times are below 10^{-9} to 10^{-11} s and most allowed transitions have lifetimes longer than this. In consequence, the line-width observed is determined by instrumental limitations and not by *uncertainty broadening.* On the other hand, typical n.m.r. spectrometers with higher resolving power ($\sim 10^8$) and lower frequencies (~ 100 MHz) have absolute resolutions of about a Hertz. This means that energy states with lifetimes less than a second give measurable line-widths.

In the absence of chemical reaction, two separate life-times can be distinguished for the upper spin state of an n.m.r. transition.† The *spin-lattice* relaxation time T_1 arises from the radiationless exchange of energy between the nucleus and its 'surroundings'. If the surroundings

† See Atkin's *Quanta* (OCS 21) for an analysis of spin relaxation processes.

are considered to contain a number of magnetic dipoles in thermal motion, their fluctuating field must occasionally be correctly phased and oriented with respect to the 'spinning' nucleus thereby permitting exchange of energy. It is interesting to note that without this mechanism energy would be absorbed until the populations of the two levels became equal and absorption would then cease. The other relaxation time T_2 arises from exchange between similar nuclei in different spin states and is known as *spin-spin* relaxation. This does not alter the net populations, in contrast to T_1, but it does affect the line-width. In fact, the *Bloch equations*, which govern n.m.r. phenomena, show that the rate of absorption is proportional to

$$T_2/[1 + 4\pi^2 T_2^2 \Delta\nu^2 + \gamma^2 H_1^2 T_1 T_2]$$

where $\Delta\nu$ is the difference in frequency from the resonance value. H_1, the intensity of the oscillating magnetic field, is usually small, so that this function can be approximated to

$$T_2/[1 + 4\pi^2 T_2^2 \Delta\nu^2].$$

This describes a *Lorentzian* line shape: it can be seen that the intensity falls to one-half its maximum value when the denominator equals 2 i.e. $\Delta\nu = 1/(2\pi T_2)$. Thus the half-width of an n.m.r. line, $1/\pi T_2$, depends solely on the spin-spin relaxation time T_2. In practice the observed half-width normally exceeds this value due to magnetic field inhomogeneity.

If the nucleus responsible for the transition is undergoing chemical reaction, it must exist in two (or more) distinct chemical environments and one can ascribe to it an effective lifetime τ similar to the spin-spin relaxation time. This also contributes to the broadening so that the total line width becomes $1/\pi T_2 + 1/\pi\tau$.

The use of this relation may be appreciated by considering the exchange reaction

$$H^{(1)}A + H^{(2)}B \rightleftharpoons H^{(2)}A + H^{(1)}B$$

in which the protons can exist in two different environments denoted by A and B. At low temperatures, where the exchange is relatively slow, two distinct lines are observed, as in Fig. 11.1a. The respective line-widths in the two environments are given by $\frac{1}{\pi}\left(\frac{1}{T_{2A}} + \frac{1}{\tau_A}\right)$ and $\frac{1}{\pi}\left(\frac{1}{T_{2B}} + \frac{1}{\tau_B}\right)$. Provided the line-widths in the absence of exchange are available, the contributions due to reaction can be measured. As the temperature is raised and exchange becomes more rapid the broadened

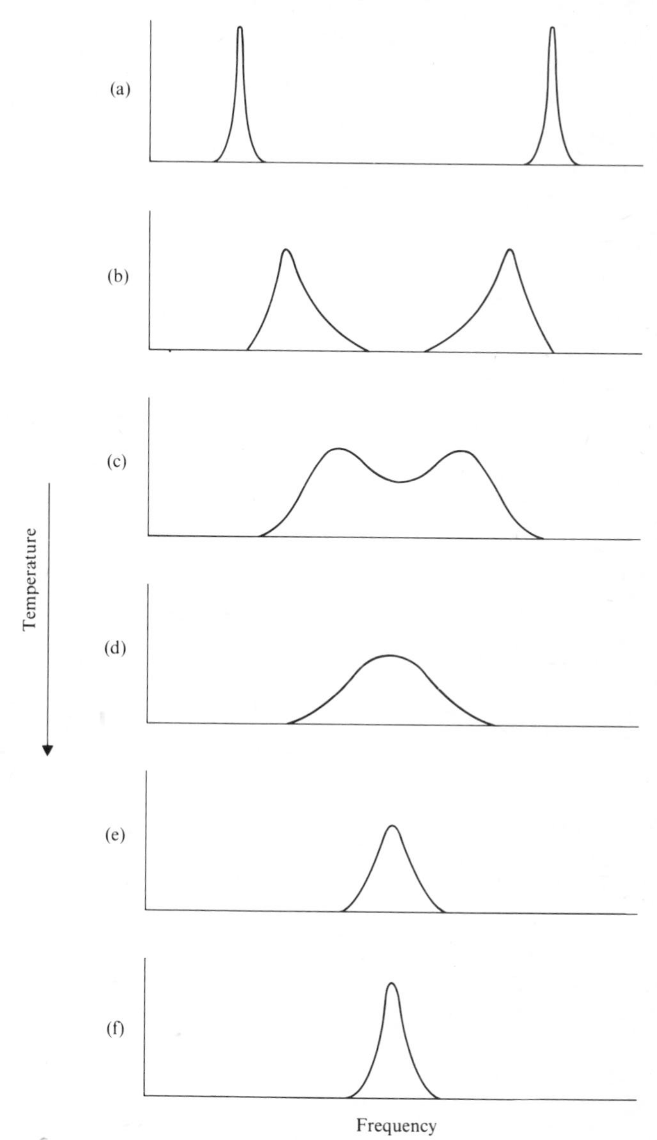

Fig. 11.1. Nuclear magnetic resonance spectra of proton exchange reaction.

lines eventually overlap and the frequency maxima move towards each other (Fig. 11.1b). Complete analysis of such spectra can be carried out using the Bloch equations but the procedure is rather complex and it is preferable to employ approximate solutions. For a simple two-state system where the lines are of equal intensity and hence characterized by a single τ value, the separation $\Delta\nu$ between the two maxima is given by

$$\frac{1}{\tau} = 2^{\frac{1}{2}}\pi({\Delta\nu_o}^2 - \Delta\nu^2)^{\frac{1}{2}},$$

where $\Delta\nu_o$ is the separation in the absence of exchange. In other situations it is usually easier to estimate values for the relaxation times and use the Bloch equations to construct an anticipated line shape, modifying the chosen τ values until the experimental shape is reproduced.

Further increase in temperature leads to coalescence of the lines, the 'dip' disappearing at $1/\tau = 2^{\frac{1}{2}}\pi\Delta\nu_o$. Beyond this point only one line is observed, which becomes narrower as the rate of exchange increases. Provided $1/\tau$ is much less than $\Delta\nu_o$, the system displays a single transition at the weighted mean frequency. In essence the exchange is too fast to be 'seen' and the ultimate line-width is governed once more by the spin-spin relaxation times T_{2A} and T_{2B}. Before this stage is attained the exchange rate can be obtained from the broadening of the single line using the relation $1/\tau \approx 4\pi(\Delta\nu_o^2/\Delta\nu)$.

The technique may be used for the investigation of a wide range of reactions although symmetrical exchange reactions of the type shown above are particularly striking examples since they are followed without the need for isotopic substitution. For practical reasons relatively large concentrations are required, say 0.1 M, so that sparingly soluble substances cannot easily be investigated, and the spectrum must be reasonably simple if the proper line widths are to be measured. Perhaps the major fundamental limitation of the technique is that the reciprocal relaxation time must not be too dissimilar from $\Delta\nu_o$, say between 0.01 $\Delta\nu_o$ and 10 $\Delta\nu_o$. This leads to typical relaxation times of 1 to 10^{-3} s for ^{1}H n.m.r. and 10^{-3} to 10^{-7} s for ^{17}O n.m.r., corresponding to first-order rate constants of 1 to $10^3\,s^{-1}$ and 10^3 to $10^7\,s^{-1}$ respectively. These rates must be achieved by appropriate adjustment of the temperature in the case of first-order reactions or, for second-order reactions, by variation of both temperature and concentration.

Double resonance

It does not always prove possible to use line shapes to measure exchange rates, either because of field inhomogeneity or of difficulty in solving the Bloch equations. In systems involving two non-equivalent sites this problem may be overcome by the *double resonance* technique.

The spectrometer is tuned to 'look at' the signal associated with one environment and the population at the other is disturbed by applying the appropriate r.f. field. As exchange proceeds, the perturbation at the second site induces a similarly-perturbed population at the site under observation. The signal intensity thus relaxes at the exchange rate.

Pulse methods

The most general technique available for the measurement of T_1 and T_2 relaxation times, and hence also for exchange rates, is the *spin-echo* method. It is only possible to give a brief description of the method here and, for a more detailed treatment, the reader is referred to standard textbooks on n.m.r.

In the presence of a static field, the magnetization vector (the resultant vector obtained by adding all the nuclear magnetic moment vectors in the sample) has an equilibrium position parallel to the field. If an r.f. pulse at the resonant frequency is applied, the magnetization vector

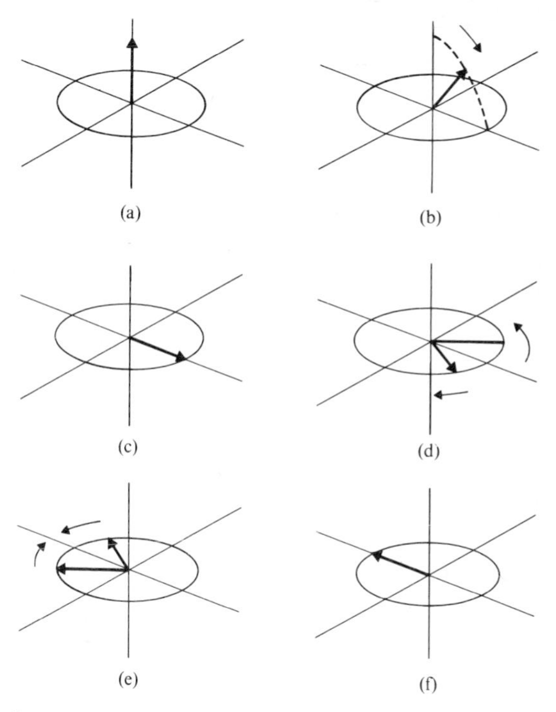

Fig. 11.2. Sequence of events in spin-echo technique. (a), (b), (c) show effect of 90° pulse on original spin vectors; (d) precession of spin vectors after termination of pulse; (e) position of vectors following application of 180° pulse; (f) 'echo'.

rotates away from the field direction into a position which depends on the pulse duration. A *90° pulse* is one which moves the vector from the z-direction into the x, y plane and a *180° pulse* reverses the direction of the vector.

In the usual method, a 90° pulse is first applied to swing the vectors into the x, y plane (see Fig. 11.2). As soon as the pulse switches off, the vectors commence precessing about the original field direction. After some time τ, a 180° pulse is applied which reverses the direction of the vectors. This leaves the vectors in the x, y plane but now moving towards each other instead of spreading out. The vectors eventually

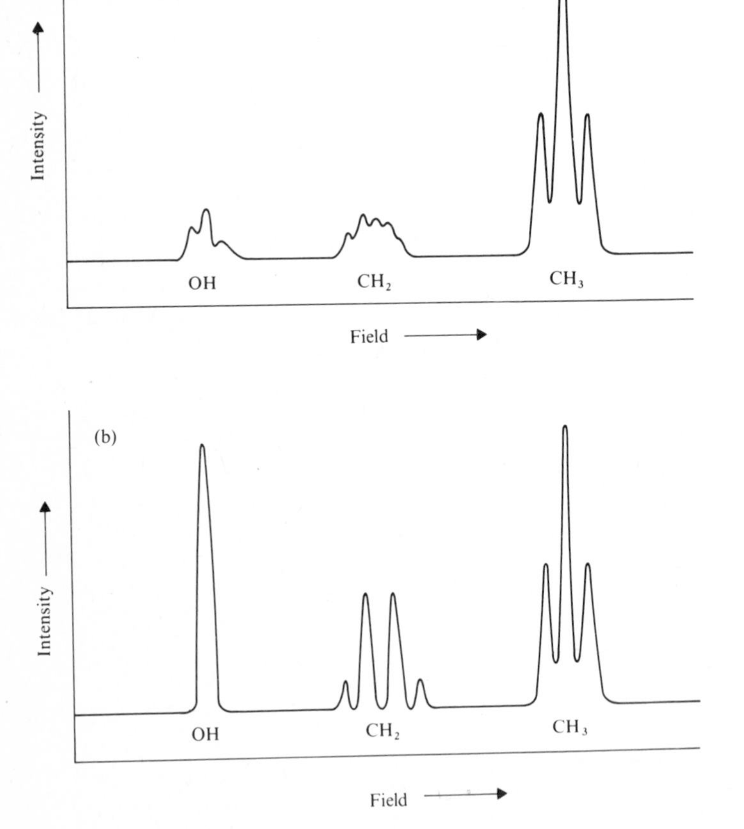

Fig. 11.3. (a) Normal n.m.r. spectrum of ethanol; (b) effect on spectrum of adding water, hydrochloric acid, or sodium hydroxide.

coincide producing a signal or *echo* at time 2τ. The magnitude of the signal will be reduced by spin relaxation processes which have occured during the time interval 2τ and the intensity thus decays according to the relation $\exp(-2\tau/T_2)$. Different pulse sequences can be employed to obtain both T_1 and T_2. Spin-echo n.m.r. instruments are available commercially which measure relaxation times varying from a few microseconds to several seconds. Among other things, the technique has been used to study hindered rotation in alkyl nitrites.

Applications

Proton-transfer reactions

Perhaps the best-known application of n.m.r. in kinetics has been for following proton-transfer reactions. The familiar ethanol spectrum shown in Fig. 11.3 provides an excellent example of the method: the features to note are the triplet form of the OH proton due to splitting by the neighbouring CH_2 group and the complex form of the CH_2 protons, which are affected both by the CH_3 and the OH protons. If water, HCl or NaOH is added, the OH triplet collapses to a single line and the CH_2 spectrum becomes a simple quartet. These changes are caused by rapid exchange of the OH proton with the additive due to processes such as

$$\mathrm{EtOH + OH^- \rightleftharpoons EtO^- + H_2O}$$

$$\mathrm{EtOH + HH'O \rightleftharpoons EtOH' + HHO}$$

$$\mathrm{EtOH + H_3O^+ \rightleftharpoons EtOH_2^+ + H_2O}$$

$$\mathrm{EtOH + EtO^- \rightleftharpoons EtO^- + EtOH}$$

$$\mathrm{EtOH + EtOH_2^+ \rightleftharpoons EtOH_2^+ + EtOH}$$

The first three processes involve water molecules and their contribution to the rate of exchange can be obtained by studying the broadening of the proton line in water. The shape of the CH_2 spectrum is affected by all five processes so that, by combining the results, one can separate out the contribution to the exchange rate associated with ethanolic species via the last two processes. By further studying the line shapes in media at various acidities it is possible to subdivide these measurements and obtain the rate constants for all five processes. This example has been chosen because of the familiarity of the ethanol spectrum but the technique has been applied to proton exchange in a wide range of compounds.

Isomerization reactions

The technique has also been applied to the kinetics of configurational changes, notably to restricted rotations about single C-C bonds in substituted ethanes and about O-N, N-N, and C-N bonds in nitrites, nitrosamines, and amides respectively. In these the major interest lies in determining the activation energies involved since they can be identified with the potential barrier heights. Ring inversions, such as those of cyclohexane and its derivatives, have been studied in this way. Such investigations tend to be more difficult than proton-transfer studies largely because of the complexity of the spectra involved.

Solvent exchange at cations

Another application of the n.m.r. technique is in the study of solvent exchange at a paramagnetic ion. Solvent molecules in the inner coordination sphere of a cation are strongly influenced by the presence of the unpaired electron and the spectrum associated with these molecules is shifted from that of the bulk solvent. In addition, the paramagnetic ion facilitates very rapid spin relaxation in these molecules. Although the spectrum of the inner sphere molecules is broad and weak, exchange between bulk solvent and inner sphere is the primary mechanism for spin relaxation of the solvent. Thus a study of line widths of the bulk solvent provides a measure of the exchange rate with the inner sphere.

Such measurements have shown that, with cations such as Mn^{2+}, Co^{2+}, Ni^{2+}, and Fe^{2+}, the rate constants are relatively low ($<10^7\ s^{-1}$) and are associated with significant activation energies (20 - 50 kJ mol^{-1}). This confirms the finding that reactions of ligands with solvated cations are frequently governed by the rate of loss of the solvent molecules.

Electron paramagnetic resonance

The use of electron paramagnetic resonance (or electron spin resonance) to monitor atom and radical concentrations in fast-flow systems has been described briefly in Chapter 7. In these examples, the e.p.r. measurements were made on steady-state systems but a very recent development has been to employ e.p.r. detection for real-time studies of flash photolytic reactions. A high time resolution (1 μs) is claimed and, combined with the considerable facility which it provides for identifying complex radicals in liquids, the technique should prove very valuable. Reaction rates can also be obtained from the spectral line widths in much the same way as with n.m.r., the major difference being that the characteristic lifetimes so obtained lie in the 10^{-5} - 10^{-9} s range in contrast to the 1 - 10^{-4} s range of proton n.m.r. Since the technique responds to paramagnetic intermediates rather than stable reactants, a simple measurement of line-width alone gives the life-time of the intermediate and hence permits determination of its reaction rate. In the presence of added reagents, line-broadening gives the rate of the electron-transfer process analogous to the proton-exchange reactions studied by n.m.r. However, the e.p.r. technique has been far less exploited for kinetic measurements than n.m.r.

Further reading

General

THE following texts together describe a very wide range of experimental techniques in considerable detail. The additional references related to the individual chapters are intended primarily to supplement the material in these general texts.

CALDIN, E. F. (1964). *Fast reactions in solution.* Blackwell, Oxford.
HAGUE, D. N. (1971). *Fast reactions.* Wiley-Interscience, London.
FRIESS, S. L., LEWIS, E. S., and WEISSBERGER, A. (1963). *Investigation of rates and mechanisms of reactions.* Vol. 8, Part 2 in the Techniques of Organic Chemistry Series (ed. A. Weissberger). Wiley-Interscience, New York.

Chapter 1

BOWDEN, F. P. and YOFFE, A. D. (1958). *Fast reactions in solids.* Butterworths, London.

Chapter 2

LEVINE, R. D. and BERNSTEIN, R. B. (1974). *Molecular reaction dynamics.* Clarendon Press, Oxford.
KINSEY, J. L. (1972). Molecular beam reactions, Chapter 6 in *Chemical kinetics* (ed. J. C. Polanyi). Vol. 9 in Physical Chemistry Series One of the MTP International Review of Science. Butterworths, London.
STEINFELD, J. I. and KINSEY, J. L. (1970). *Prog. react. kinet.* **1,** 1.
TOENNIES, J. P. (1968). *Ber.* (*dtsch.*) *Bunsenges. phys. Chem.* **72,** 927.

Chapter 3

CARRINGTON, T. and POLANYI, J. C. (1972). Chemiluminescent reactions, Chapter 5 in *Chemical kinetics* (ed. J. C. Polanyi). Vol. 9 in Physical Chemistry Series One of the MTP International Review of Science. Butterworths, London.
KOMPA, K. L., PARKER, J. H., and PIMENTEL, G. C. (1968). *J. chem. Phys.* **49,** 4257. (chemical laser method)
PARKER, J. H. and PIMENTEL, G. C. (1969). *J. chem. Phys.* **51,** 91. (equal-gain method)

Chapter 4

NORRISH, R. G. W. (1965). *Chem. Br.* **1,** 289.
PORTER, G. (1967). Flash photolysis and primary processes in the excited state, in *Nobel Symposium 5: fast reactions and primary processes in chemical kinetics* (ed. S. G. Claesson). Interscience, New York.
PORTER, G. (1967). Flash photolysis, Chapter 5 in *Photochemistry and reaction kinetics* (ed. P. G. Ashmore, F. S. Dainton, and T. M. Sugden). Cambridge University Press, Cambridge.
THRUSH, B. A. (1967). Flash photolytic studies of free radicals in the gas phase, Chapter 6 in *Photochemistry and reaction kinetics* (ed. P. G. Ashmore, F. S. Dainton, and T. M. Sugden). Cambridge University Press, Cambridge.
DORFMAN, L. M. and MATHESON, M. S. (1965). *Prog. react. kinet.* **3,** 237 (pulse radiolysis).

DAINTON, F. S. (1967). The chemistry of the electron, in *Nobel Symposium 5: fast reactions and primary processes in chemical kinetics* (ed. S. G. Claesson). Interscience, New York.

Chapter 5

BRADLEY, J. N. (1962). *Shock waves in chemistry and physics.* Methuen and Wiley, London.

BRADLEY, J. N. (1963). *R. Inst. Chem. (London), Lect. Ser.* **6**, 1.

BRADLEY, J. N. (1967). *Advmt. Sci. Lond.* **23**, 585.

GAYDON, A. G. and HURLE, I. R. (1963). *The shock tube in high temperature chemical physics.* Chapman and Hall, London.

GREENE, E. F. and TOENNIES, J. P. (1964). *Chemical reactions in shock waves.* Edward Arnold, London.

BELFORD, R. L. and STREHLOW, R. A. (1969). *A. Rev. phys. Chem.* **20**, 247.

Chapter 6

PHILLIPS, L. F. (1973). *Prog. react. kinet.* **7**, 83.

Chapter 7

DENBIGH, K. G. and PAGE, F. M. (1954). *Discuss. Faraday Soc.* **17**, 145. (stirred-flow reactor)

KAUFMAN, F. (1961). *Prog. react. kinet.* **1**, 1. (gas-phase flow systems)

WESTENBERG, A. A. (1973). *Prog. react. Kinet.* **7**, 23. (electron spin resonance detection)

BRADLEY, J. N. (1969). *Flame and combustion phenomena.* Methuen, London.

GAYDON, A. G. and WOLFHARD, H. G. (1970). *Flames: their structure, radiation and temperature* (3rd ed.). Chapman and Hall, London.

Chapter 8

CROSS, J. E. (1972). Relaxation techniques, Chapter 10 in *Chemical kinetics* (ed. J. C. Polanyi). Vol. 9 in Physical Chemistry Series One of the MTP International Review of Science. Butterworths, London.

Chapter 9

MATHESON, A. J. (1971). *Molecular acoustics.* Wiley-Interscience, London. (ultrasonics)

BLOCK, H. and NORTH, A. M. (1970). *Adv. molec. Relax. Processes* **1**, 309. (dielectric relaxation)

DAVIES, M. (1965). *Some electrical and optical aspects of molecular behaviour.* Pergamon, London. (dielectric relaxations)

HILL, N. E., VAUGHAN, W. E., PRICE, A. H. and DAVIES, M. (1969). *Dielectric properties and molecular behaviour.* Van Nostrand-Reinhold, London. (dielectric relaxation)

NORTH, A. M. (1973). *Essays Chem.* **4**, 1.

Chapter 10

DELAHAY, P. (1954). *New instrumental methods in electrochemistry.* Interscience, London.

Chapter 11

EMSLEY, J. W., FEENEY, J. and SUTCLIFFE, L. H. (1965). *High resolution nuclear magnetic resonance spectroscopy.* Pergamon, Oxford.
POPLE, J. A., SCHNEIDER, W. G. and BERNSTEIN, H. J. (1959). *High-resolution nuclear magnetic resonance.* McGraw-Hill, New York.

Index